T0191506

Springer Theses

Recognizing Outstanding Ph.D. Research

For further volumes:
http://www.springer.com/series/8790

Aims and Scope

The series "Springer Theses" brings together a selection of the very best Ph.D. theses from around the world and across the physical sciences. Nominated and endorsed by two recognized specialists, each published volume has been selected for its scientific excellence and the high impact of its contents for the pertinent field of research. For greater accessibility to non-specialists, the published versions include an extended introduction, as well as a foreword by the student's supervisor explaining the special relevance of the work for the field. As a whole, the series will provide a valuable resource both for newcomers to the research fields described, and for other scientists seeking detailed background information on special questions. Finally, it provides an accredited documentation of the valuable contributions made by today's younger generation of scientists.

Theses are accepted into the series by invited nomination only and must fulfill all of the following criteria

- They must be written in good English.
- The topic should fall within the confines of Chemistry, Physics, Earth Sciences, Engineering and related interdisciplinary fields such as Materials, Nanoscience, Chemical Engineering, Complex Systems and Biophysics.
- The work reported in the thesis must represent a significant scientific advance.
- If the thesis includes previously published material, permission to reproduce this must be gained from the respective copyright holder.
- They must have been examined and passed during the 12 months prior to nomination.
- Each thesis should include a foreword by the supervisor outlining the significance of its content.
- The theses should have a clearly defined structure including an introduction accessible to scientists not expert in that particular field.

Ali Kakhbod

Resource Allocation in Decentralized Systems with Strategic Agents

An Implementation Theory Approach

Doctoral Thesis accepted by
the University of Michigan, MI, USA

Author
Dr. Ali Kakhbod
Electrical and System Engineering
University of Pennsylvania
Philadelphia
USA

Supervisor
Prof. Demosthenis Teneketzis
Electrical Engineering and Computer Science
University of Michigan
Ann Arbor
USA

ISSN 2190-5053 ISSN 2190-5061 (electronic)
ISBN 978-1-4899-8614-6 ISBN 978-1-4614-6319-1 (eBook)
DOI 10.1007/978-1-4614-6319-1
Springer New York Heidelberg Dordrecht London

© Springer Science+Business Media New York 2013
Softcover re-print of the Hardcover 1st edition 2013
This work is subject to copyright. All rights are reserved by the Publisher, whether the whole or part of the material is concerned, specifically the rights of translation, reprinting, reuse of illustrations, recitation, broadcasting, reproduction on microfilms or in any other physical way, and transmission or information storage and retrieval, electronic adaptation, computer software, or by similar or dissimilar methodology now known or hereafter developed. Exempted from this legal reservation are brief excerpts in connection with reviews or scholarly analysis or material supplied specifically for the purpose of being entered and executed on a computer system, for exclusive use by the purchaser of the work. Duplication of this publication or parts thereof is permitted only under the provisions of the Copyright Law of the Publisher's location, in its current version, and permission for use must always be obtained from Springer. Permissions for use may be obtained through RightsLink at the Copyright Clearance Center. Violations are liable to prosecution under the respective Copyright Law.
The use of general descriptive names, registered names, trademarks, service marks, etc. in this publication does not imply, even in the absence of a specific statement, that such names are exempt from the relevant protective laws and regulations and therefore free for general use.
While the advice and information in this book are believed to be true and accurate at the date of publication, neither the authors nor the editors nor the publisher can accept any legal responsibility for any errors or omissions that may be made. The publisher makes no warranty, express or implied, with respect to the material contained herein.

Printed on acid-free paper

Springer is part of Springer Science+Business Media (www.springer.com)

Parts of this thesis have been published in the following journal articles:

A. Kakhbod and D. Teneketzis. An Efficient Game Form for Unicast Service Provisioning. *IEEE Transactions on Automatic Control* (*TAC*). vol. **57**, no. 2, 392–404, 2012.

A. Kakhbod and D. Teneketzis. Power Allocation and Spectrum Sharing in Multi-User, Multi-Channel Systems with Strategic Users. *IEEE Transactions on Automatic Control* (*TAC*). vol. **57**, no. 9, 2338–2342, 2012.

A. Kakhbod and D. Teneketzis. An Efficient Game Form for Multi-rate Multicast Service Provisioning. *IEEE Journal on Selected Areas in Communications* (*JSAC*), *Special Issue on Economics of Communication Networks and Systems*. vol. **30**, no. 11, 2093–2104, 2012.

To my family

Supervisor's Foreword

Ali Kakhbod's thesis is a significant contribution to decentralized resource allocation in communication networks (wired and wireless) with strategic users. Kakhbod investigated three classes of problems. (C1) Unicast service provisioning in wired networks. (C2) Multi-rate multicast service provisioning in wired networks. (C3) Power allocation and spectrum sharing in multi-user multi-channel wireless communication systems. In all these problems the network's users are strategic. Problems in (C1) are market problems; problems in (C2) are a combination of markets and public goods; problems in (C3) are public goods. For all three classes of problems Kakhbod discovered decentralized resource allocation mechanisms that possess the following properties. (P1) They implement in Nash equilibria the social welfare maximizing correspondence or the Pareto correspondence. (P2) They are budget balanced. (P3) They are individually rational. Within the context of communication networks, no other allocation mechanism available in the literature possesses all properties (P1)–(P3). Furthermore, Kakhbod's thesis contributes to the state of the art of mechanism design. To the best of my knowledge, research on mechanism design never addressed problems that are a combination of markets and public goods. Kakhbod's results on multi-rate multicast are a significant contribution (the first contribution) to this important class of problems.

Ann Arbor Demosthenis Teneketzis

Acknowledgments

It is my pleasure to thank the many people who made this thesis possible. I would like to sincerely thank my advisor Professor Demosthenis Teneketzis for his continuous support and encouragement, for the opportunity he gave me to conduct independent research, and for his exemplary respect. I still remember the day when he had told me, "You will have to believe me about the potential of this field". Today when I see the vast scope of research in information economics and its potential to create an impact on a huge number of applications, I do not have enough words to thank him for introducing me to this field. His vision, patience, and dedication to educate his students enabled me to pursue my research in a field that I could have never imagined before the graduate school. I express my humble gratitude to him for his friendly and ever inspiring mentorship.

I would also like to thank my dissertation committee members, Professor Achilleas Anastasopoulos, Professor Mingyan Liu, Professor Yusufcan Masatlioglu, and Professor Asuman Ozdaglar for kindly accepting to be on my dissertation committee. Their interest in my research has been a great source of pleasure and motivation for me. I am especially thankful to Professor Asuman Ozdaglar for her valuable comments, in particular about the content of Chap. 2.

I would like to express my sincere gratitude to my former teachers. In particular, I would like to thank Professor Tilman Borgers and Professor Lones Smith for giving me extra time for discussions and providing me valuable insights on information economics.

I would like to thank my friends who made my life in Ann Arbor enjoyable and sociable, especially my great roommate, Mohammad Fallahi, for all I learned from him and all the couch-sitting sessions! and also Ali Nazari and Aria Ghasemian Sahebi for their invaluable friendship. In addition, I would like to thank Ashutosh Nayyar and Shrutivandana Sharma for many discussions and helpful suggestions. Specially, I am thankful to Demos and Barbara for many evenings I have spent at their home.

Last, but not least, I am grateful to my parents, my brother, and my sister for their endless love and support and for putting my happiness before their own. I would like to dedicate this thesis to them.

Contents

Abstract

In this thesis we present an implementation theory approach to decentralized resource allocation problems with strategic agents in communication networks. For wired networks we study the unicast and multi-rate multicast service provisioning problem. For wireless networks we study the problem of power allocation and spectrum sharing where each user's transmissions create interference to all (or subset of) network users, and each user has only partial information about the network. We investigate these problems under the implementation theory scenario where agents/users are strategic, self-utility maximizing. We present key concepts and ideas from implementation theory that are relevant to the problems. We formulate the unicast service provisioning problem as a market allocation problem, the power allocation and spectrum sharing problem as a public goods allocation problem, and the multi-rate multicast service provisioning problem as the combination of a market and a public goods allocation problems. For each problem, we develop a game form that (i) implements in Nash equilibria the optimal allocations of the corresponding centralized problem; (ii) is individually rational; and (iii) results in budget balance at all Nash equilibria and feasible off-equilibrium.

Chapter 1
Introduction

1.1 Motivation

Networks exist in a vast variety of real world systems. They have played an important role in the social and technological growth of our society. Some prominent examples of networked systems are urban and transportation systems, military systems, political/social networks, production and consumer markets, supply-chains, energy markets, internet, web data centers, electronic commerce systems, sensor networks and telecommunication systems. Because of the diversity of network applications, networks are studied in a wide range of professional and academic domains including engineering, business management and social science.

Irrespective of the diversity of applications, a fundamental similarity in all of the above networks is that, (i) the network consists of multiple agents that interact with and influence each other; (ii) each agent has different characteristics and a different individual role in the network; and (iii) the actions of individual agents together with their interactions determine the function/performance of the network.[1]

Apart from the fundamental similarities in the structure and function of networks described by the above mentioned features, an identical objective in their design is their efficient operation. This requires optimization of network performance measures. As mentioned above, a network's performance is determined by the collective actions of network agents. Actions that are critical in determining a network performance are consumption/generation of resources by network agents and their decisions regarding network tasks. Therefore, for a network to achieve its performance objective, proper allocation of the network's resources and coordination of the network agents' decisions are extremely important. With the technological and social advancement, many networks such as the internet, energy markets and e-commerce systems are expanding at a very fast pace. The resources that are required for the operation of these networks, e.g. bandwidth, fossil fuels, and web server resources

[1] An alternative term for networks that captures the above characteristics is multi-agent systems. In many applications such as electronic commerce, artificial intelligence and social networks, the use of the term multi-agent system is more common.

A. Kakhbod, *Resource Allocation in Decentralized Systems with Strategic Agents*, Springer Theses, DOI: 10.1007/978-1-4614-6319-1_1,
© Springer Science+Business Media New York 2013

often do not increase at the same rate. Therefore, in these cases resource allocation and utilization become even more crucial for efficient network operation.

In the context of communication networks some of the important resources are bandwidth, energy, coding schemes, relay routes, and the physical space available for the network. The important performance measures are data communication rate, probability of error, communication delay, battery life, interference, mobility of agents, ability to dynamically adjust to varying channel conditions, etc. The requirement to efficiently utilize the above resources to achieve desirable performance gives rise to several challenging resource allocation problems in communication networks. Examples of such problems are spectrum/rate/code allocation that govern throughput and delay, power and code allocation that govern interference and battery life, admission control that governs the number of agents in the network, topology control that governs the placement and interconnections of network agents, and dynamic resource allocation that looks at the above aspects under dynamic situations. The numerous applications and the technical challenge of these resource allocation problems make them an important and exciting area of communication networks research. This motivated us to investigate some of these problems in this thesis.

1.2 Key Issues and Challenges in Resource Allocation

The challenge in resource allocation comes from: (1) the fact that the network is an informationally decentralized system; (2) the network's users/agents may behave strategically (i.e. they may behave selfishly). Networks are informationally decentralized systems. Each user's utility is its own private information. Users are unaware of each others' utilities and of the resources (e.g. bandwidth, buffers, spectrum) available to the network. The network (network manager) knows the network's topology and its resources but is unaware of the users' utilities. If information were centralized, the resource allocation problem could be formulated and solved as a mathematical programming problem or as a dynamic programming problem. Since information is not centralized such formulations are not possible. The users' strategic behavior along with the decentralization of information lead to problems that are conceptually difficult and computationally formidable.

The challenge is: (1) To determine a message exchange process among the network and users, and an allocation rule (based on the outcome of the message exchange process) that eventually lead to a resource allocation that is optimal for the centralized problem. (2) To take into account, in the determination of the allocation mechanism, the possible strategic (selfish) behavior of the networks users.

As is evident from the above, addressing decentralized resource allocation problems requires a framework that can provide a systematic methodology for the design of decentralized resource allocation mechanisms by harnessing the decentralized information characteristics of the networks and the behavioral characteristics of the agents. One such framework for the systematic study of decentralized resource allocation problems with strategic agents is provided by implementation theory which is a branch of mathematical economics.

1.3 Contribution of the Thesis

In this thesis we investigate decentralized resource allocation problems with strategic users. The main contributions of the thesis are:

- The formulation and solution of decentralized resource allocation problems that arise in communication networks (wired and wireless) within the context of implementation theory.
- The construction and analysis of game forms for decentralized resource allocation problems that have not been previously investigated within the context of implementation theory.

Thus, this thesis contributes not only to the solution of important technological problems but also to the state of the art of implementation theory.

We have investigated three classes of decentralized resource allocation problems motivated by communication networks. (i) Market problems; (ii) public goods problems; and (iii) problems that are a combination of markets and public goods. For all these classes of problems we developed game forms/mechanisms which have the following desirable properties. (P1) They implement in Nash equilibria (NE) either the social welfare maximizing correspondence or the Pareto correspondence. That is, the allocations corresponding to all NE of the game induced by the game form/mechanism are either optimal solutions of the corresponding centralized resource allocation problem or Pareto optimal; conversely, to every optimal solution of the centralized resource allocation problem there corresponds a NE of the game induced by the mechanism. (P2) They are individually rational. That is, users voluntarily participate in the allocation process, as their utility at the allocations corresponding to all NE of the game induced by the mechanism is larger than the utility they obtain by not participating in the allocation process. (P3) They are budget balanced at the allocations corresponding to all NE as well as all feasible allocations corresponding to off equilibrium strategies.

We proceed now to describe our contribution to each of the aforementioned problems.

In Chap. 3 we address the unicast service provisioning problem with strategic users that arises in wired networks. This is a market problem. When the users' utilities are concave, we propose a mechanism that possesses properties (P2), (P3) and implements in NE the social welfare maximizing correspondence. When the users' utilities are quasi-concave, the allocations corresponding to all NE of the game induced by the mechanism are Pareto optimal. The game form proposed in this thesis for unicast service provisioning problem with strategic users is currently the only existing mechanism that possesses properties (P1)–(P3).

In Chap. 4 we investigate power allocation and spectrum sharing in multi-user multi-channel systems arising in wireless networks. This is a public goods problem. We propose a mechanism that possesses properties (P2), (P3) and implements in NE the Pareto correspondence. The game form proposed for this problem is currently the only existing mechanism that possesses properties (P1)–(P3). Two key features of our problem formulation are: (F1) The allocation space is discrete. (F2) There are no assumptions about concavity, monotonicity, or quasi-linearity of the users' utility functions. Decentralized resource allocation problems with features (F1)–(F2) have not been previously investigated within the context of implementation theory. Thus, the results of Chap. 4 are a contribution to the state of the art of implementation theory.

In Chap. 5 we address the multi-rate multicast service provisioning problem. This is the combination of a market and a public goods problem. We present a mechanism which possesses properties (P2), (P3) and implements in NE the social welfare maximizing correspondence. To the best of our knowledge this work is the first to address multi-rate multicast service provisioning with strategic users; in all previously existing literature on multi-rate multicast service provisioning, users were assumed to be non-strategic. Problems that are a combination of a market and public goods were not previously investigated within the context of implementation theory. Thus, the results of Chap. 5 are also a contribution to the state of the art of implementation theory.

1.4 Organization of the Thesis

This thesis is organized as follows: In Chap. 2 we present a brief introduction to the key ideas and results of implementation theory that are relevant to the problems we investigate in this thesis. In Chap. 3 we present the unicast service provisioning problem with strategic agents arising in wired networks. In Chap. 4 we present the problem of power allocation and spectrum sharing in multi-user, multi-channel systems with strategic agents arising in wireless networks. In Chap. 5 we present the multi-rate multicast service provisioning problem with strategic agents arising in wired networks. We conclude in Chap. 6.

Chapter 2
Implementation Theory

In this section we present key ideas and results from implementation theory that are relevant to the topics of the thesis.

2.1 What is Implementation Theory?

2.1.1 Preliminaries

Implementation theory is a component of mechanism design. It provides an analytical framework for situations where resources have to be allocated among agents/users but the information needed to make these allocation decisions is dispersed and privately held, and the agents/users possessing the private information behave strategically and are self-utility maximizers. In any situation where the information needed to make decisions is dispersed, it is necessary to have information exchange among the agents/users possessing the information. Allocation decisions are made after the information exchange process terminates. Implementation theory provides a systematic methodology for designing an information exchange process followed by an allocation rule that leads to allocation decisions that are "optimal" with respect to some pre-specified performance metric.

The objectives of implementation theory are:

(1) To determine, for any given performance metric, whether or not there exists an information exchange process and an allocation rule that achieve optimal allocations with respect to that metric when the users possess private information and are strategic.
(2) To determine systematic methodologies for designing information exchange processes and allocation rules that achieve optimal allocations with respect to performance metrics for which the answer to (1) is positive.

A. Kakhbod, *Resource Allocation in Decentralized Systems with Strategic Agents*, Springer Theses, DOI: 10.1007/978-1-4614-6319-1_2,
© Springer Science+Business Media New York 2013

(3) To determine alternative criteria for the design of information exchange processes and allocation rules that lead to "satisfactory" allocations for situations where the answer to (1) is negative.

The key concept in the development of implementation theory is that of game form or mechanism. A game form/mechanism consists of two components: (1) A message/strategy space, that is, a communication alphabet through which the agents/users exchange information with one another. (2) An allocation rule (called outcome function) that determines the allocations after the communication and information exchange process terminates. Most mechanisms employ monetary incentives and payments to achieve desirable resource allocations. In such cases, the outcome function specifies the resource allocations as well as the monetary incentives and payments.

A game form along with the agents'/users' utilities defines a game. The allocations made (through the outcome function) at the equilibria of the game determine the result of the decentralized allocation problem. The key objectives in the design of a game form/mechanism are:

1. To provide incentives to the strategic agents/users so that they prefer to participate in the allocation process rather than abstain from it.
2. To obtain, at all equilibria of the game induced by the mechanism, allocations that are optimal with respect to some pre-specified performance metric (criterion). For example, it may be desirable that the allocations obtained by the game form/mechanism are the same as those obtained by the solution of the corresponding centralized allocation problem.
3. To obtain a balanced budget at all equilibria of the game induced by the mechanism. That is, at all equilibria, the money received by some of the system's agents/users as part of the incentives provided by the mechanism must be equal to the money paid by the rest of the agents/users.

2.1.2 Game Forms/Mechanisms

In the implementation theory/mechanism design framework, a centralized resource allocation problem is described by the triple $(\mathcal{E}, \mathcal{A}, \pi)$: the environment space $\mathcal{E}$, the action/allocation space $\mathcal{A}$ and the goal correspondence/social choice correspondence/social choice rule π. Below, we briefly describe each component separately. Let $\mathcal{N} = \{1, 2, \ldots, N\}$ be the set of agents/users.

Environment Space ($\mathcal{E}$): We define the environment space $\mathcal{E}$ of an allocation problem to be the set of individual preferences (or the set of utilities), endowments and the technology taken together. The environment $\mathcal{E}$ is the set of circumstances that can not be changed either by the designer of the allocation mechanism or by the agents/users that participate in the allocation mechanism.

The environment space $\mathcal{E}$ is the cartesian product of the users' individual environment spaces $\mathcal{E}_i$, i.e., $\mathcal{E} := \mathcal{E}_1 \times \mathcal{E}_2 \times \cdots \times \mathcal{E}_N$. A realization $e \in \mathcal{E}$ of the environment is a collection of the users' individual realizations $e_i, e_i \in \mathcal{E}_i, i = 1, 2, \ldots, N$, that is, $e = (e_1, e_2, \ldots, e_N)$.

Action Space ($\mathcal{A}$): We define the action space $\mathcal{A}$ of a resource allocation problem to be the set of all possible actions/resource allocations.

Goal Correspondence/Social Choice Rule/Social Choice Correspondence (π): Goal correspondence is a map from $\mathcal{E}$ to $\mathcal{A}$ which assigns to every environment $e \in \mathcal{E}$ the set of actions/allocations which are solutions to the centralized resource allocation problem associated with/ corresponding to the decentralized resource allocation problem under consideration. That is,

$$\pi : \mathcal{E} \to \mathcal{A}.$$

The setting described above corresponds to the case where one of the agents (e.g. a network manager) has enough information about the environment so as to determine the actions according to the goal correspondence π. Generally this is not the case. Usually, different agents have different information about the environment. For this reason it is desired to devise a mechanism for information exchange and resource allocation that leads, for every instance e of the resource allocation problem, to an allocation in $\pi(e)$.

When the system's agents are strategic the resource allocation mechanism is described by a N-user/agent game form $(\mathcal{M}, f)$, where $\mathcal{M} = \Pi_{i=1}^{N} \mathcal{M}_i$ is the message space, specifying for each user i, $i = 1, 2, \ldots, N$, the set of messages $\mathcal{M}_i$ that user/agent i can communicate to other users, and f is an outcome function that describes the actions that are taken for every $m := (m_1, m_2, \ldots, m_N) \in \mathcal{M}$; that is

$$f : \mathcal{M} \to \mathcal{A}.$$

The game form $(\mathcal{M}, f)$ is common knowledge [1, 2] among all the N agents/users. Note that a game form is different from a game, as the consequence of a profile m of messages is an allocation (or a set of allocations if f is a correspondence) rather than a vector of utility payoffs. Once a realization $e \in \mathcal{E}$ of the environment is specified, a game form induces a game.

Within the context of implementation theory, a decentralized resource allocation process proceeds in three steps:

1. The mechanism designer announces the game form $(\mathcal{M}, f)$.
2. An instance $e \in \mathcal{E}$ of the environment is realized. The realization of environment e specifies, among other things, the utilities $u_i, i \in \mathcal{N}$, of all agents. Depending on its utilities and the specified mechanism, each agent decides whether or not to participate in the mechanism. The agents that choose not to participate in the allocation process get some exogenously specified "reservation utility", which is usually a number independent of the environment e; we set this number to be zero.
3. The agents who choose to participate in the allocation process play the game induced by the mechanism. In this game, $\mathcal{M}_i$ is the strategy space of player i,

and for every strategy profile $m \in \mathcal{M}$, $u_i(f(m))$ is the utility payoff of player i. We denote this game by $(\mathcal{M}, f, e)$.

The mechanism designer is interested in the outcomes that occur at the equilibria of the game induced by the game form.

2.1.3 Implementation in an Appropriate Equilibrium Concept

A solution/equilibrium concept specifies the strategic behavior of the agents/users faced with a game $(\mathcal{M}, f, e)$ induced by the game form $(\mathcal{M}, f)$. Consequently, an equilibrium concept is a correspondence Λ that identifies a subset of $\mathcal{M}$ for any given specification $(\mathcal{M}, f, e)$. We define for every environment $e \in \mathcal{E}$,

$$\mathcal{A}_\Lambda(m, f, e) := \{a \in \mathcal{A} : \exists m \in \Lambda(\mathcal{M}, f, e) : f(m) = a\} \tag{2.1}$$

as the set of outcomes associated with the solution concept Λ, when the environment is e.

The solution/equilibrium concept appropriate for a decentralized resource allocation problem depends on the information that is available to the agents/users about the environment. For example, if agent i, $i \in \mathcal{N}$, knows $e_i \in \mathcal{E}_i$ and has a probability mass function on $\mathcal{E}_{-i} = \Pi_{j=1, j \neq i}^{N} \mathcal{E}_j$, then an appropriate solution concept is a Baysian Nash equilibrium (BNE) [3]. On the other hand, if agent i, $i \in \mathcal{N}$ knows $e_i \in \mathcal{E}_i$, and $\mathcal{E}_j$, for all $j \neq i$, then an appropriate solution concept is a Nash equilibrium (NE) [4], or a sub-game perfect NE or a sequential NE [5].

Definition 2.1 A goal correspondence $\pi : \mathcal{E} \mapsto \mathcal{A}$ is implemented (respectively, weakly implemented) by the game form $(\mathcal{M}, f)$ in the equilibrium concept Λ if $\mathcal{A}_\Lambda(\mathcal{M}, f, e) = \pi(e)$ (respectively, $\mathcal{A}_\Lambda(\mathcal{M}, f, e) \subset \pi(e)$) for all $e \in \mathcal{E}$.

Definition 2.2 A goal correspondence $\pi : \mathcal{E} \rightarrow \mathcal{A}$ is said to be implementable (respectively, weakly implementable) in solution/equilibrium concept Λ if there exists a game form $(\mathcal{M}, f)$ that implements (respectively, weakly implements) it.

Within the context of implementation theory there have been significant developments in the characterization of goal correspondences that can be implemented in the following solution concepts: dominant strategies [6, 7]; Nash equilibria [8–11]; refined Nash equilibria, such as sub-game perfect equilibria [12, 13], undominated Nash equilibria [14–17], trembling hand perfect Nash equilibria [18]; Bayesian Nash equilibria [19–22].

2.1.4 Implementation in Nash Equilibrium and Maskin's Mechanism

In the problems investigated in this thesis we consider Nash equilibrium (NE) as the solution/equilibrium concept. For any $(\mathcal{M}, f, e)$, a pure NE is a message/strategy profile $m^* := (m_1^*, m_2^*, \dots, m_N^*) \in \mathcal{M}$ such that for all $i \in \mathcal{N}$,

$$u_i(f(m_i^*, m_{-i}^*)) \geq u_i(f(m_i, m_{-i}^*)), \tag{2.2}$$

for all $m_i \in \mathcal{M}_i$, where $m_{-i}^* := (m_1^*, m_2^*, \dots, m_{i-1}^*, m_{i+1}^*, \dots, m_N^*)$ and $u_i, i \in \mathcal{N}$, are the utility functions of the agents under the realization e of the environment. Let $NE(\mathcal{M}, f, e)$ be the set of NE of the game $(\mathcal{M}, f, e)$ and

$$\mathcal{A}_{NE} := \mathcal{A}_{NE}(\mathcal{M}, f, e) := \{a \in \mathcal{A} | \exists m \in NE(\mathcal{M}, f, e) \text{ s.t. } f(m) = a\}. \tag{2.3}$$

The game form $(\mathcal{M}, f)$ implements (respectively, weakly implements) a social choice rule π in Nash equilibrium if

$$\mathcal{A}_{NE}(\mathcal{M}, f, e) = \pi(e)$$

(respectively, $\mathcal{A}_{NE}(\mathcal{M}, f, e) \subset \pi(e)$) for all $e \in \mathcal{E}$.

In his seminal paper [4], Maskin characterized social choice rules that can be implemented in NE, and constructed a mechanism that achieves implementation in NE. To state and discuss the main result in [4] we need to define the concepts of weak no-veto power and monotonicity.

Definition 2.3 A goal correspondence/social choice rule $\pi : \mathcal{E} \to \mathcal{A}$ satisfies weak no-veto power if for any $e \in \mathcal{E}$, any outcome $a \in \mathcal{A}$ that is the top ranked alternative of at least $N - 1$ agents under the given environment e (that is, a simultaneously maximizes individual utilities of at least $N - 1$ agents) belongs to $\pi(e)$.

In words, a social choice rule satisfies weak no-veto power if, whenever all agents except possibly one agree that an alternative is *top-ranked*, (i.e. no other outcome is higher in their preference orderings), then that alternative is in the social choice set; the remaining agent can not veto it.

Definition 2.4 A social choice rule $\pi : \mathcal{E} \to \mathcal{A}$ is *monotonic* if for all $e := (e_1, e_2, \dots, e_N)$, $\hat{e} := (\hat{e}_1, \hat{e}_2, \dots, \hat{e}_N)$ and $a \in \mathcal{A}$, $a \in \pi(\hat{e}_1, \hat{e}_2, \dots, \hat{e}_N)$ whenever:

(i) $a \in \pi(e_1, e_2, \dots, e_N)$,
(ii) for all $b \in \mathcal{A}$ and $i \in \mathcal{N}$, $u_i(a) \geq u_i(b)$ implies $\hat{u}_i(a) \geq \hat{u}_i(b)$,

where $u_i, \hat{u}_i$ are utilities of agent i under e_i and $\hat{e}_i$, respectively.

In words, monotonicity of π says the following: Suppose that under a profile of utility functions $u_1, u_2, \dots, u_N$, $u_i \in e_i$, $\forall\, i \in \mathcal{N}$, the outcome a is in the choice set $\pi(e)$. Furthermore, suppose that the environment e is altered to $\hat{e}$ so that under the new profile $\hat{u} := (\hat{u}_1, \hat{u}_2, \dots, \hat{u}_N)$, $\hat{u}_i \in \hat{e}_i$ for all $i \in \mathcal{N}$, the outcome a does not fall in any agent's preference ordering relative to any outcome in $\mathcal{A}$. Then, the outcome a must be in the choice set $\pi(\hat{e})$. Monotonicity is satisfied by many social choice rules including the "social welfare maximizing correspondence" and the "Pareto correspondence" [8, 23–25].[1]

We can now state Maskin's fundamental result on Nash implementation.

Theorem 2.5 ([4]) *If a social choice rule* $\pi : \mathcal{E} \to \mathcal{A}$ *is implementable in NE then it is monotonic. Furthermore, if the number of agents is at least* 3 *and* π *is monotonic and satisfies the weak no-veto power condition, then* π *is implementable in NE.*

The proof of the above theorem is constructive. Given a social choice rule π that satisfies monotonicity and weak no-veto power, Maskin constructs a game form that implements π. We present the construction of the game form in words. Such a presentation reveals the complexity of the mechanism and provides a justification as to why we pursue alternative approaches in this thesis. Before we proceed with the description of the game form we need to define the lower contour set of u_i, $i \in \mathcal{N}$, at outcome $a \in \mathcal{A}$.

Definition 2.6 For each $a \in \mathcal{A}$ and $u_i \in \mathcal{U}_i$, $i \in \mathcal{N}$, let

$$\mathcal{LC}(a, u_i) := \{b \in \mathcal{A} \text{ s.t. } u_i(a) \geq u_i(b)\} \tag{2.4}$$

$\mathcal{LC}(a, u_i)$ is the lower contour set of u_i at $a \in \mathcal{A}$.

In words, the lower contour set of u_i at $a \in \mathcal{A}$ is the set of outcomes in $\mathcal{A}$ that someone with utility function u_i does not prefer to a.

We now proceed with the description of Maskin's game form. The message space for each agent $i \in \mathcal{N}$ is $\mathcal{M}_i := \mathcal{U}_1 \times \mathcal{U}_2 \times \cdots \times \mathcal{U}_N \times \mathcal{A} \times \mathbb{Z}_{++}$, where $\mathbb{Z}_{++}$ is the space of positive integers. That is, each agent's message/strategy is a profile of the agents' utilities, an outcome $a \in \mathcal{A}$, and a positive integer. The outcome function is defined as follows:

(i) If all agents announce the same message/strategy $m_i = (u_1, u_2, \ldots, u_N, a, K)$, $i \in \mathcal{N}$, and $a \in \pi(e_1, e_2, \ldots, e_N)$, $(u_i \in e_i \; \forall i \in \mathcal{N})$, then $f(m_1, m_2, \ldots, m_N) = a$. That is, if all agents are unanimous in their strategy, and their proposed outcome is in the choice set $\pi(e)$, then the outcome is a.

(ii) Suppose all agents $j \neq i$ announce the same strategy $m_j = (u_1, u_2, \ldots, u_N, a, K)$ and $a \in \pi(e_1, e_2, \ldots, e_N)$. Let $m_i = (u'_1, u'_2, \ldots, u'_N, a', K')$ be the ith agent's message/strategy. Then,

$$f(m_1, m_2, \ldots, m_N) = \begin{cases} a' & \text{if } a' \in \mathcal{LC}(a, u_i) \\ a & \text{if } a' \notin \mathcal{LC}(a, u_i). \end{cases} \tag{2.5}$$

That is, suppose all players but one propose the same strategy and the proposed outcome $a \in \pi(e_1, e_2, \ldots, e_N)$. Then, agent i, the odd-agent out, gets his pro-

[1] In Chaps. 3 and 5, the social choice rule/goal correspondence is the social welfare maximizing correspondence and in Chap. 4 the goal correspondence is the Pareto correspondence.

posal a', provided that a' is in the lower contour set of a of the ordering that the other agents proposed for him. Otherwise, the outcome is a.

(iii) If neither (i) nor (ii) applies, then $f(m_1, m_2, \ldots, m_N) = a^i$, where $i := \max\{i \mid k^i = \max_{j \in \mathcal{N}} k^j\}$. In words, when neither (i) nor (ii) applies, the outcome is the one proposed by the agent that has the highest index among those whose proposed integer is maximal.

Maskin proved in [4] that the above-described game form/mechanism implements π in NE.

From the above description it is clear that Maskin's mechanism requires, in general, an infinite dimensional message space. That is why in this thesis we do not follow Maskin's approach. We follow a different approach, outlined in Chaps. 3–5, which requires a finite dimensional message space and a particular interpretation of NE. Below we present this interpretation of NE.

2.1.5 Interpreting Nash Equilibrium

We present two interpretations of Nash equilibrium which appear in Nash's original work [24]. The first is the "mass-action" interpretation of NE points. According to this interpretation, it is unnecessary to assume that agents participating in the game have full knowledge of the structure of the game, or the ability to go through any complex reasoning process. But it is assumed that the participants have the ability to accumulate empirical information, obtained through repeated plays of the game and to evaluate, using this empirical information, the relative advantage of the various pure strategies they have at their disposal. The evaluation of empirical information determines, as the number of repeated plays of the game increases, the agents' NE strategies. Quoting Nash,

> *It is unnecessary to assume that participants have full knowledge of the total structure of the game... but the participants are supposed to accumulate empirical information on the relative advantages of the various pure strategies at their proposal*, J. Nash, PhD thesis ([24] p. 21).

Implicit in this interpretation of NE is the assumption that the game's environment e is stable, that is, it does not change before the agents reach their equilibrium strategies. Nash's "mass-action" interpretation of NE has also been adopted by Reichelstein and Reiter [26], and Groves and Ledyard [27]. The authors of [26, 27] consider resource allocation problems with strategic agents who have private information, adopt NE as the solution concept and state,

> *We interpret our analysis as applying to an unspecified (message exchange) process in which users grope their way to a stationary message and in which the Nash property is a necessary condition for stationarity*, Reichelstein and Reiter ([26] p. 664).

and,

> *We do not suggest that each agent knows e when he computes m, ... We do suggest, however, that the 'complete information' Nash equilibrium game-theoretic equilibrium messages may be the possible equilibrium of the iterative process—that is, the stationary messages—just as the demand–equal–supply price is thought of the equilibrium of some unspecified market dynamic process*, Groves and Ledyard ([27] pp. 69–70).

In the second interpretation of NE, it is assumed that the agents know the full structure of the game in order to be able to predict the equilibrium strategies. This interpretation of NE is rationalistic and idealizing.

In this thesis, we will adopt the "mass-action" interpretation of NE. In the problems we investigate in Chaps. 3 and 5 the environment is stable (as the network where the network is wireless, the environment is assumed to be stable (i.e. it does not change) during the allocation process.

2.1.6 Desirable Properties of Game Forms

In addition to implementation in an appropriate equilibrium concept, the mechanism designer should try to achieve the other objectives mentioned in Sect. 2.1.1. We formally define the properties of a mechanism associated with those objectives in this section.

2.1.6.1 Individual Rationality

One of the objectives in the design of a game form is to incentivize all the agents to voluntarily participate in the allocation process under any possible environment. Consider any environment $e \in \mathcal{E}$. If under e, agent i decides not to participate, its overall utility is zero (see Sect. 2.1.2). If agent i decides to participate in the game induced by the mechanism, its utility is $u_i(f(m^*))$ where m^* is an equilibrium of the game $(\mathcal{M}, f, e)$ induced by the mechanism. Under $e \in \mathcal{E}$, an agent participates in the allocation process if at all equilibria m^* of the game $(\mathcal{M}, f, e)$, $u_i(f(m^*)) \geq 0$. We can now define individually rational mechanisms as follows:

Definition 2.7 A mechanism/game form $(\mathcal{M}, f)$ is individually rational if for all $e \in \mathcal{E}$, for all equilibria m^* of the game $(\mathcal{M}, f, e)$ and for all $i \in \mathcal{N}$, $u_i(f(m^*)) \geq 0$, where u_i is the utility function of agent i in the environment e, and 0 is the reservation utility a user receives if it decides not to participate in the allocation process (cf. Sect. 2.1.2).

2.1.6.2 Budget Balance

Strategic agents are often incentivized to follow the rules of the mechanism through monetary tax or subsidy. Some agents are induced to accept allocations that may not be their most preferred ones (under the realization e of the environment) by receiving money (subsidy). Conversely, some agents are induced to pay money (tax) for receiving their most preferred allocations. It is desirable that for any environment $e \in \mathcal{E}$, at every equilibrium of the game $(\mathcal{M}, f, e)$ the sum of taxes paid by some agents should be equal to the sum of subsidies received by the rest of the agents. Any mechanism $(\mathcal{M}, f)$ that possesses the above property is said to be *budget balanced at equilibrium*. Budget balance is also desirable at all out of equilibrium messages that result in feasible allocations (i.e. allocations that satisfy the problem's constraints) for the following practical reason. Suppose the mechanism designer specifies, along with the mechanism, an iterative message exchange process (tâtonnement process) which for any environment $e \in \mathcal{E}$ is guaranteed to converge to an equilibrium of the game induced by the mechanism. In practice, the message exchange process may terminate when it reaches sufficiently close to the equilibrium (but not the equilibrium). If the mechanism is not budget balanced at these out of equilibrium terminal messages, then possible large amounts of unclaimed money may be left.

References

1. Aumann RJ (1976) Agreeing to disagree. Ann Stat 4:1236–1239
2. Washburn R, Teneketzis D (1984) Asymptotic agreement among communicating decision makers. Stochastics 13:103–129
3. Palfrey T, Srivastava S (1993) Bayesian implementation: fundamentals of pure and applied economics 53. Harwood Academic, New York
4. Maskin E (1999) Nash equilibrium and welfare optimality. Rev Econ Stud 66(1):23–38
5. Myerson R (1981) Optimal auction design. Math Oper Res 6(1):58–73
6. Dasgupta P, Hammond P, Maskin E (1979) The implementation of social choice rules: some general results on incentive compatibility. Rev Econ Stud 46(2):185–216
7. Green J, Laffont JJ (1979) Incentives in public decision making. North-Holland, Amsterdam
8. Maskin E (1985) The theory of implementation in Nash equilibrium: a survey. In: Hurwicz L, Schmeidler D, Sonnenschein H (eds) Social goals and social organization. Cambridge University Press, Cambridge, pp 173–204
9. Maskin E, Sjőstrőm T (2002) Implementation theory. In: Arrow K, Sen A, Suzumura K (eds) Handbook of social choice and welfare. North Holland, Amsterdam
10. Saijo T (1988) Strategy space reduction in maskin's theorem: sufficient conditions for nash implementation. Econometrica 56(3):693–700
11. Wang Q, Peha J, Sirbu M (1997) Optimal pricing for integrated-services networks. In: McKnight LW, Bailey JP (eds) Internet economics, 3rd edn. MIT Press, Cambridge, pp 353–376
12. Abreu D, Sen A (1990) Subgame perfect implementation: a necessary and almost sufficient condition. J Econ Theory 50:285–299
13. Moore J, Repullo R (1990) Nash implementation: a full characterization. Econometrica 58(5):1083–1099
14. Abreu D, Matsushima H (1992) Virtual implementation in iteratively undominated strategies: complete information. Econometrica 60(5):993–1008
15. Jackson M (1992) Implementation of undominated strategies. Rev Econ Stud 59(4):757–775
16. Jackson M, Palfrey T, Srivastava S (1994) Undominated nash implementation in bounded mechanisms. Games Econ Behav 6(3):474–501
17. Palfrey T, Srivastava S (1991) Nash implementation using undominated strategies. Econometrica 59(2):479–502

18. Sjostrom T (1993) Implementation in perfect equilibrium. Soc Choice Welf 10:97–106
19. Jackson MO (1991) Bayesian implementation. Econometrica 59(1):461–477
20. Palfrey T, Srivastava S (1989) Implementation with incomplete information in exchange. Econometrica 57(1):115–134
21. Palfrey T, Srivastava S (1992) Implementation in bayesian equilibrium: the multiple equilibrium problem in mechanism design. In: Laffont J (ed) Advances in economic theory, in econometric society monographs. Cambridge University Press, New York
22. Postlewaite A, Schmeideler D (1986) Implementation in differential information economies. J Econ Theory 39(1):14–33
23. Mas-Colell A, Whinston MD, Green JR (2005) Microeconomic theory. Oxford University Press, New York
24. Hurwicz L, Reiter S (2006) Designing economic mechanisms. Cambridge University Press, New York
25. Jackson M (2001) A crash course in implementation theory. Soc Choice Welf 18(4):655–708
26. Reichelstein S, Reiter S (1988) Game forms with minimal strategy space. Econometrica 56(3):661–692
27. Groves T, Ledyard J, (1987) Incentive compatibility since, (1972) In: Groves T, Radner R, Reiter S (eds) Information, incentives, and economic mechanisms: essays in honor of Leonid Hurwicz. University of Minnesota Press, Minneapolis, pp 48–109

Chapter 3
Unicast Service Provisioning

3.1 Introduction

Most of today's networks, called integrated services networks support the delivery of a variety of services to their users each with its own quality of service (QoS) requirements (e.g. delay, percentage of packet loss, jitter, etc.). As the number of services offered by the network and the demand for the services increase, the need for efficient network operation increases. One of the key factors that contributes to efficient network operation is the efficient utilization of the network's resources.

In this chapter we investigate the unicast service provisioning problem in wired networks with arbitrary topology and strategic users. We formulate the problem as a market allocation with strategic users. The key issues and challenges associated with this problem have been discussed in Sect. 1.2 of the thesis. Here we propose a game form/mechanism for the solution of the problem, we analyze the mechanism's properties and compare our results to the existing literature on unicast service provisioning with strategic users.

3.1.1 Contribution of the Chapter

We investigate the unicast service provisioning problem in wired networks with arbitrary topology and strategic users. The main contribution of this chapter of the thesis is the discovery of a decentralized rate allocation mechanism for unicast service provisioning in networks with arbitrary/general topology and strategic users, which possesses the following properties.
When each user's utility is concave, then:

(P1) The mechanism implements the solution of the centralized unicast service provisioning problem in Nash equilibria.

A. Kakhbod, *Resource Allocation in Decentralized Systems with Strategic Agents*, Springer Theses, DOI: 10.1007/978-1-4614-6319-1_3,
© Springer Science+Business Media New York 2013

(P2) The mechanism is individually rational, that is, the network users voluntarily participate in the rate allocation process.

(P3) The mechanism is budget-balanced at all the feasible allocations (cf. Sect. 2.1.6.2), that is, at all the allocations that correspond to NE messages/strategies as well as at all the feasible allocations that correspond to off-equilibrium messages/strategies.

When each user's utility is quasi-concave but differentiable, then:

The mechanism possesses properties (P2) and (P3).

(P4) Every NE of the game induced by the mechanism results in a Walrasian equilibrium ([1] Ch. 15), consequently, a Pareto optimal allocation.

To the best of our knowledge, none of the decentralized resource allocation mechanisms proposed so far for the unicast service provisioning problem in communication networks possesses simultaneously all three properties (P1–P3) when the network's topology is general/arbitrary, the users are strategic and their utilities are concave. Furthermore, we are not aware of the existence of any publications in unicast service provisioning containing the analysis of a decentralized rate allocation mechanism when the users are strategic and their utilities are quasi-concave.

We now compare in more detail our contributions with the existing literature.

3.1.2 Comparison with Related Work

Recently, within the context of communication networks, researchers have investigated decentralized resource allocation problems under the assumption that users behave strategically (i.e. they are not price-takers, they do not necessarily obey the rules of the mechanism but have to be induced to follow the rules). Within the context of wired networks, decentralized resource allocation mechanisms have been proposed and analyzed in [2–15].

We now explain why the proposed mechanism and the above results are distinctly different from all game forms/mechanisms proposed so far for the unicast service provisioning problem with strategic users.

Most of the previous work on the unicast service provisioning problem in networks with general topology is based on Vickrey-Clark-Groves(VCG)-type mechanisms, [9–11, 15–19]. The game forms/mechanisms proposed in [15] and [9] induce games that establish the existence of a unique Nash equilibrium at which the allocation is globally optimal under some conditions; but these mechanisms are not budget-balanced even at equilibrium. The mechanisms/game forms proposed in [10, 11, 16] induce games that have multiple NE; these mechanisms are not budget-balanced even at equilibrium, and the allocations corresponding to the Nash equilibria are not always globally optimal (that is these mechanisms do not implement in Nash equilibria the solution of the centralized unicast service provisioning problem). Our mechanism is not of the VCG-type, thus, it is philosophically different from those of [9–11, 15, 16].

The work in [12, 13] and [20] deals with single link networks. For these single-link networks the authors of [13] proposed a class of efficient (optimal) allocation mechanisms, called ESPA, for the allocation of a single divisible good. ESPA mechanisms were further developed in [12]. It is not currently known whether ESPA mechanisms implement in Nash equilibria the optimal solution of the unicast service provisioning problem in networks with arbitrary/general topology. The network model considered in this chapter has arbitrary/general topology.

In [7, 8] the authors show that when the resource allocation mechanism proposed in [5] is considered under the assumption that the users are strategic and NE is the equilibrium concept, the allocations corresponding to any NE are different from any allocations that are optimal solutions of the corresponding centralized unicast service provisioning problem; that is, the allocation corresponding to any NE suffer from a certain efficiency loss. Particularly, in [8] it is shown that there exists a lower bound on the efficiency loss. The mechanism we propose in this chapter is distinctly different from those of [7, 8]. Our mechanism results in the same performance as optimal centralized allocations, that is, the allocations corresponding to any NE of the game induced by our mechanism are efficient.

Philosophically, our work is most closely related to [14], but it is distinctly different from [14] for the following reasons: (1) The game form proposed in this chapter is distinctly different from that of [14]. (2) The mechanism of [14] is not balanced off equilibrium. (3) In the mechanism of [14] there is no coupling among the games that are being played at different links. In our mechanism such a coupling exists (see Sect. 3.3), and results in a balanced-budget off equilibrium.

Finally, we are not aware of any publication, other than the mechanism we propose in this chapter, containing the analysis of a decentralized rate allocation mechanism for unicast service provisioning when the users are strategic and their utilities are quasi-concave.

3.1.3 Organization of the Chapter

The rest of the chapter is organized as follows. In Sect. 3.2 we formulate the unicast service provisioning problem with strategic users. In Sect. 3.3 we describe the allocation mechanism/game form we propose for the solution of the unicast service provisioning problem. In Sect. 3.4 we analyze the properties of the proposed mechanism. In Sect. 3.5 we discuss how the game form/mechanism presented in this chapter can be implemented in a network. In Sect. 3.6 we investigate the properties of the game form proposed in this chapter when the users' utilities $u_i, i \in \mathcal{N}$, are quasi-concave. The proofs of all the results established in this chapter appear in Appendix A.

3.2 The Unicast Problem with Strategic Network Users, Problem Formulation

In this section we present the formulation of the unicast problem in wired communication networks with strategic users. We proceed as follows, In Sect. 3.2.1 we formulate the centralized unicast service provisioning problem the solution of which we want to implement in Nash equilibria. In Sect. 3.2.2 we formulate the decentralized unicast service provisioning problem with strategic network users and, state our assumptions and our objective.

3.2.1 The Centralized Problem

We consider a wired network with N, $N > 3$, users. The set of these users is denoted by $\mathcal{N}$, i.e. $\mathcal{N} = \{1, 2, \ldots, N\}$. The network topology, the capacity of the network links, and the routes assigned to users' services are fixed and given. The users' utility functions have the form

$$\mathcal{V}_i(x_i, t_i) = u_i(x_i) - t_i, \quad i = 1, 2, \ldots, N. \tag{3.1}$$

The term $u_i(x_i)$ expresses user i's *satisfaction* from the service x_i it receives. The term t_i represents the *tax* (money) user i pays for the services it receives. We assume that u_i is a concave and increasing function of the service x_i user i receives, and $t_i \in \mathbb{R}$. When $t_i > 0$ user i pays money for the services it receives; this money is paid to other network users. When $t_i < 0$ user i receives money from other users. Overall, the amount of money paid by some of the network users must be equal to the amount of money received by the rest of the users so that $\sum_{i \in \mathcal{N}} t_i = 0$. Denote by $\mathbf{L}$ the set of links of the network, by c_l the capacity of link l, and by $\mathcal{R}_i$ the set of links l, $l \in \mathbf{L}$, that form the route of user i, $i = 1, 2, \ldots, N$ (as pointed out above each user's route is fixed). We assume that a central authority (the network manager) has access to all of the above information. The objective of this authority is to solve the following centralized optimization problem that we call **Max**.

$$\textbf{Max} \qquad \max_{x_i} \sum_{i=1}^{N} u_i(x_i) \tag{3.2}$$

subject to

$$\sum_{i: l \in \mathcal{R}_i} x_i \leq c_l, \quad \forall\, l \in \mathbf{L}, \tag{3.3}$$

$$x_i \geq 0, \quad \forall i \in \mathcal{N}, \tag{3.4}$$

$$\sum_{i=1}^{N} t_i = 0, \; t_i \in \mathbb{R}, \quad \forall i \in \mathcal{N}. \tag{3.5}$$

The inequalities in (3.3) express the capacity constraints that must be satisfied at each network link. The inequalities in (3.4) express the fact that the users' received services $x_i, i \in \mathcal{N}$ must be nonnegative. The equality in (3.5) express the fact that the budget must be balanced, i.e. the total amount of money paid by some of the users must be equal to the amount of money received by the rest of the users.

Let $\mathcal{U}$ denote the set of functions

$$u : \mathbb{R}_+ \cup \{0\} \to \mathbb{R}_+ \cup \{0\}$$

where u is concave and increasing. Let $\mathbf{T}$ denote the set of all possible network topologies, network resources and user routes. Consider problem **Max** for all possible realizations

$$(u_1, \ldots, u_N, T) \in \mathcal{U}^N \times \mathbf{T}$$

of the users' utilities, the network topology, its resources and the users' routes. Then, the solution of **Max** for each $(u_1, u_2 \cdots, u_N, T) \in \mathcal{U}^N \times \mathbf{T}$ defines a map

$$\pi : \mathcal{U}^N \times \mathbf{T} \to \mathcal{A}$$

where $\mathcal{A} \in \mathbb{R}_+^n \times \mathbb{R}^N$ is the set of all possible rate/bandwidth allocations to the network's users and the taxes (resp. subsidies) paid (resp. received) by the users. We call π the solution of the centralized unicast service provisioning problem.

3.2.2 The Decentralized Problem with Strategic Users

We consider the network model of the previous section with the following assumptions on its information structure.

(A1) Each user knows only his own utility; this utility is his own private information.

(A2) Each user behaves strategically, that is, each user is not a price-taker. The users's objective is to maximize its own utility function.

(A3) The network manager knows the topology and resources of the network. This knowledge is the manager's private information. The network manager is not a profit-maker (i.e. he does not have a utility function).

(A4) The network manager receives requests for service from the network users. Based on these requests, he announces to each user $i, i \in \mathcal{N}$:

  (i) The set of links that form user i's route, $\mathcal{R}_i$; that is, the network manager chooses the route for each user and this route remains fixed throughout the user's service.

  (ii) The capacity of each link in $\mathcal{R}_i$.

(A5) Based on the network manager's announcement, each strategic user competes for resources (bandwidth) at each link of his route with the other users in that link.[1]

From the above description it is clear that the information in the network is decentralized. Every user knows his own utility but does not know the other users' utilities or the network's topology and its resources. The network manager knows the network's topology and its resources, but does not know the users' utilities. It is also clear that the network manager (which is not profit maker) acts like an accountant who sets up the users' routes, specifies the users competing for resources/bandwidth at each link, collects the money from the users i that pay tax (i.e. $t_i > 0$) and distributes it to those users j that receive money (i.e. $t_j < 0$).

As a consequence of assumptions (A1–A5) we have at each link of the network a decentralized resource allocation problem which can be studied/analyzed within the context of implementation theory. These decentralized resource allocation problems are not independent/decoupled, as the rate that each user receives at any link of his own route must be the same. This constraint is dictated by the nature of the unicast service provisioning problem and has a direct implication on the nature of the mechanism/game form we present in Sect. 3.3.

Under the above assumptions the objective is to determine a game form/mechanism which has the following properties,

(P1) It implements in NE the social welfare maximizing correspondence defined by the centralized problem **Max**. (Note that the social welfare maximizing correspondence is implementable in NE, cf. Sect. 2.1.4).
(P2) It is individually rational, that is, for every realization

$$(u_1, u_2, \ldots, u_N, T) \in \mathcal{U}^N \times \mathbf{T},$$

the network users voluntarily participate in the bandwidth allocation process.
(P3) For every realization $(u_1, u_2, \ldots, u_N, T) \in \mathcal{U}^N \times \mathbf{T}$ it is budget balanced at every NE of the game it induces, as well as at all off equilibrium messages that result in feasible allocations.

In the following two sections we present a mechanism/game form for the problem formulated in this section and prove that it possess properties (P1–P3) stated above.

3.3 A Mechanism for Rate Allocation

In Sect. 3.3.1, we specify a mechanism/game form for the decentralized rate allocation problem formulated in Sect. 3.2. In Sect. 3.3.2, we discuss and interpret the components of the mechanism.

[1] During the play of the game at each link $l \in \mathbf{L}$, each user of link l learns the set of the other users competing for bandwidth at l.

3.3.1 Specification of the Mechanism

A game form/mechanism (cf. Sect. 2.1.2) consists of two components $\mathcal{M}$ and f. The component $\mathcal{M}$ denotes the users' *message/strategy space*. The component f is the *outcome function*; f defines for every message/strategy profile, the bandwidth/rate allocated to each user and the tax (subsidy) each user pays (receives).

For the decentralized resource allocation problem formulated in Sect. 3.2 we propose a game form/mechanism the components of which we describe below.

Message space: The message/strategy space for user i, $i = 1, 2, ..., N$, is given by $\mathcal{M}_i \subset R_+^{|\mathcal{R}_i|+1}$. Specifically, a message of user i is of the form

$$m_i = (x_i, p_i^{l_{i_1}}, p_i^{l_{i_2}}, \cdots, p_i^{l_{i_{|\mathcal{R}_i|}}})$$

where $0 \le x_i \le \min_{l \in \mathcal{R}_i} c_l$ and $0 \le p_i^{l_{i_k}} \le M, k = 1, 2, , \ldots, |\mathcal{R}_i|, 0 < M < \infty$, M is large, and $|\mathcal{R}_i|$ denotes the number of links along route $\mathcal{R}_i, i \in \mathcal{N}$. The component x_i denotes the bandwidth/rate user i requests at all the links of his route. The component $p_i^{l_{i_j}}, j = 1, 2, \ldots, |\mathcal{R}_i|$, denotes the price per unit of bandwidth user i is willing to pay at link l_{i_j} of his route.

As noted in Sect. 3.2.2, the nature of the unicast service provisioning problem dictates/requires that the bandwidth/rate allocated to any user $i, i \in \mathcal{N}$, must be the same at all links of his route. Thus, the nature of message m_i is a consequence of the above requirement.

Outcome Function: The outcome function f is given by

$$f : \mathcal{M}_1 \times \mathcal{M}_2 \times \cdots \times \mathcal{M}_N \to (\mathbb{R}_+^N \times \mathbb{R} \times \mathbb{R} \cdots \times \mathbb{R})$$

and is defined as follows. For any $m := (m_1, m_2, \ldots, m_N) \in \mathcal{M} := \mathcal{M}_1 \times \mathcal{M}_2 \times \cdots \times \mathcal{M}_N$,

$$f(m) = f(m_1, m_2, \cdots, m_N) = (x_1, x_2, \ldots, x_N, t_1, t_2, \cdots, t_N)$$

where $x_i, i \in \mathcal{N}$, is the amount of bandwidth/rate allocated to user i (this is equal to the amount of bandwidth user $i, i \in \mathcal{N}$, requests), and $t_i, i \in \mathcal{N}$, is determined by t_i^l, the tax (subsidy) user i pays (receives) for link $l, l \in \mathcal{R}_i$, and by other additional subsidies $\mathcal{Q}^i$ that user i may receive. We proceed now to specify $t_i^l, l \in \mathcal{R}_i$, and $\mathcal{Q}^i$ for every user $i \in \mathcal{N}$.

The tax $t_i^{l_{i_j}}, j = 1, 2, \ldots, |\mathcal{R}_i|, i \in \mathcal{N}$, is defined according to the number of users using link l. Let $\mathcal{G}^l$ denotes the set of users using link l and let $|\mathcal{G}^l|$ denote the cardinality of $\mathcal{G}^l$. We consider three cases[2]

[2] We consider only the cases where $|\mathcal{G}^l| \ge 2$. If $|\mathcal{G}^l| = 1$ and $i \in \mathcal{G}^l$, then $t_i^l = 0 \cdot 1\{x_i \le c_l\} + \frac{1\{x_i > c_l\}}{1 - 1\{x_i > c_l\}}$.

- CASE 1, $|\mathcal{G}^l| = 2$
  Let $i,\ j \in \mathcal{G}^l$. Then,

$$t_i^l = p_j^l x_i + \frac{(p_i^l - p_j^l)^2}{\alpha} - 2p_j^l(p_i^l - p_j^l)\left(\frac{x_i + x_j - c^l}{\gamma}\right) + \frac{1\{x_i > 0\}1\{x_i + x_j - c^l > 0\}}{1 - 1\{x_i > 0\}1\{x_i + x_j - c^l > 0\}} \tag{3.6}$$

$$t_j^l = p_i^l x_j^l + \frac{(p_j^l - p_i^l)^2}{\alpha} - 2p_i^l(p_j^l - p_i^l)\left(\frac{x_i + x_j - c^l}{\gamma}\right) + \frac{1\{x_j > 0\}1\{x_i + x_j - c^l > 0\}}{1 - 1\{x_j > 0\}1\{x_i + x_j - c^l > 0\}} \tag{3.7}$$

where α and γ are positive constants that are sufficiently large and, the function $1\{A\}$, used throughout the chapter, is defined as follows

$$1\{A\} = \begin{cases} 1 - \epsilon & \text{if } A \text{ holds;} \\ 0 & \text{otherwise.} \end{cases}$$

where ϵ is bigger than zero and sufficiently small[3]; ϵ is chosen by the mechanism designer.

- CASE 2, $|\mathcal{G}^l| = 3$
  Let $i,\ j$ and $k \in \mathcal{G}^l$. Then

$$t_i^l = P_{-i}^l x_i + (p_i^l - P_{-i}^l)^2 - 2P_{-i}^l(p_i^l - P_{-i}^l)\left(\frac{\mathcal{E}_{-i}^l + x_i}{\gamma}\right) + \frac{1\{x_i > 0\}1\{x_i + x_j + x_k - c^l > 0\}}{1 - 1\{x_i > 0\}1\{x_i + x_j + x_k - c^l > 0\}} + \Omega_i^l \tag{3.8}$$

$$t_j^l = P_{-j}^l x_j + (p_j^l - P_{-j}^l)^2 - 2P_{-j}^l(p_j^l - P_{-j}^l)\left(\frac{\mathcal{E}_{-j}^l + x_j}{\gamma}\right) + \frac{1\{x_j > 0\}1\{x_i + x_j + x_k - c^l > 0\}}{1 - 1\{x_j > 0\}1\{x_i + x_j + x_k - c^l > 0\}} + \Omega_j^l \tag{3.9}$$

$$t_k^l = P_{-k}^l x_k + (p_k^l - P_{-k}^l)^2 - 2P_{-k}^l(p_k^l - P_{-k}^l)\left(\frac{\mathcal{E}_{-k}^l + x_k}{\gamma}\right) + \frac{1\{x_k > 0\}1\{x_i + x_j + x_k - c^l > 0\}}{1 - 1\{x_k > 0\}1\{x_i + x_j + x_k - c^l > 0\}} + \Omega_k^l \tag{3.10}$$

[3] Therefore, when A and B (both) hold, then $\frac{1\{A\}1\{B\}}{1-1\{A\}1\{B\}} \approx \frac{1}{0^+}$ is well defined and it becomes a large number.

where,

$$\begin{aligned}
P_{-i}^{l} &= \frac{p_j^l + p_k^l}{2}, P_{-j}^{l} = \frac{p_k^l + p_i^l}{2}, P_{-k}^{l} = \frac{p_j^l + p_i^l}{2}, \\
\mathcal{E}_{-i}^{l} &= x_j + x_k - c_l, \mathcal{E}_{-j}^{l} = x_i + x_k - c_l, \mathcal{E}_{-k}^{l} = x_i + x_j - c_l \\
\mathcal{E}_i^l &= 2x_i - c_l, \mathcal{E}_j^l = 2x_j - c_l, \mathcal{E}_k^l = 2x_k - c_l,
\end{aligned} \tag{3.11}$$

and Ω_i^l is defined as

$$\begin{aligned}
\Omega_i^l =& \frac{\sum_{\substack{r \in \mathcal{G}^l \\ r \neq i}} \sum_{\substack{s \in \mathcal{G}^l \\ s \neq i,r}} \left(2p_r^l p_s^l (1 + \frac{x_r}{\gamma}) - x_r p_s^l\right)}{(\left|\mathcal{G}^l\right| - 1)(\left|\mathcal{G}^l\right| - 2)} - \frac{\sum_{\substack{j \in \mathcal{G}^l \\ r \neq i}} p_r^{l\,2}}{\left|\mathcal{G}^l\right| - 1} - P_{-i}^{l\,2} \\
& - 2\frac{\mathcal{E}_{-i}^{l} P_{-i}^{l\,2}}{\gamma}.
\end{aligned} \tag{3.12}$$

The terms Ω_j^l and Ω_k^l are defined in a way similar to Ω_i^l.

- CASE 3, $\left|\mathcal{G}^l\right| > 3$
  Let $i \in \mathcal{G}^l \subseteq \mathcal{N}$. Then,

$$\begin{aligned}
t_i^l =& P_{-i}^{l} x_i + (p_i^l - P_{-i}^{l})^2 - 2P_{-i}^{l}(p_i^l - P_{-i}^{l})\left(\frac{\mathcal{E}_{-i}^{l} + x_i}{\gamma}\right) \\
& + \frac{1\{x_i > 0\}1\{\mathcal{E}_{-i}^{l} + x_i > 0\}}{1 - 1\{x_i > 0\}1\{\mathcal{E}_{-i}^{l} + x_i > 0\}} + \Phi_i^l
\end{aligned} \tag{3.13}$$

where,

$$P_{-i}^{l} = \frac{\sum_{\substack{j \in \mathcal{G}^l \\ j \neq i}} p_j^l}{\left|\mathcal{G}^l\right| - 1}, \quad \mathcal{E}_{-i}^{l} = \sum_{\substack{j \in \mathcal{G}^l \\ j \neq i}} x_j - c^l, \quad \mathcal{E}_i^l = (\left|\mathcal{G}^l\right| - 1)x_i - c^l,$$

and

$$\begin{aligned}
\Phi_i^l =& \frac{\sum_{\substack{j \in \mathcal{G}^l \\ j \neq i}} \sum_{\substack{k \in \mathcal{G}^l \\ k \neq i,j}} \left(2p_j^l p_k^l (1 + \frac{x_j}{\gamma}) - x_j p_k^l\right)}{(\left|\mathcal{G}^l\right| - 1)(\left|\mathcal{G}^l\right| - 2)} \\
& + \frac{\sum_{\substack{j \in \mathcal{G}^l \\ j \neq i}} \sum_{\substack{k \in \mathcal{G}^l \\ k \neq i,j}} \sum_{\substack{r \in \mathcal{G}^l \\ r \neq i,j,k}} 2p_k^l (p_j^l \mathcal{E}_r^l - x_j p_r^l)}{\gamma(\left|\mathcal{G}^l\right| - 1)^2(\left|\mathcal{G}^l\right| - 3)}
\end{aligned}$$

$$+\frac{\sum_{\substack{j\in\mathcal{G}^l\\ j\neq i}}\sum_{\substack{k\in\mathcal{G}^l\\ k\neq i,j}}2p_k^l(p_j^l\mathcal{E}_k^l-x_jp_k^l)}{\gamma(|\mathcal{G}^l|-1)^2(|\mathcal{G}^l|-2)}$$
$$-\frac{\sum_{\substack{j\in\mathcal{G}^l\\ j\neq i}}p_j^{l\,2}}{|\mathcal{G}^l|-1}-P_{-i}^{l\,2}-2\frac{\mathcal{E}_{-i}^lP_{-i}^{l\,2}}{\gamma}. \tag{3.14}$$

Next we specify additional subsidies $\mathcal{Q}^i$ that user i, $i\in\mathcal{N}$, may receive. For that matter we consider all links $l\in\mathbf{L}$ such that $|\mathcal{G}^l|=2$ or $|\mathcal{G}^l|=3$. For each link l, with $|\mathcal{G}^l|=2$ we define the quantity

$$\begin{aligned}\mathcal{Q}^{\{l:|\mathcal{G}^l|=2\}}:&=-2\frac{(p_i^l-p_j^l)^2}{\alpha}-p_j^lx_i-p_i^l+x_j\\ &\times\left[2p_j^l(p_i^l-p_j^l)+2p_i^l(p_j^l-p_i^l)\right]\left(\frac{x_i+x_j-c_l}{\gamma}\right)\\ &=o(1)-p_j^lx_i-p_i^lx_j;\end{aligned} \tag{3.15}$$

for each link with $|\mathcal{G}^l|=3$ we define

$$\begin{aligned}\mathcal{Q}^{\{l:|\mathcal{G}^l|=3\}}:=&\frac{1}{\gamma}\left(-2P_{-i}^{l\;2}x_{-i}+2P_{-i}^lp_i^l\mathcal{E}_{-i}^l-2P_{-j}^{l\;2}x_{-j}\right)\\ &+\frac{1}{\gamma}\left(2P_{-j}^lp_j^l\mathcal{E}_{-j}^l-2P_{-k}^{l\;2}x_{-k}+2P_{-k}^lp_k^l\mathcal{E}_{-k}^l\right).\end{aligned} \tag{3.16}$$

Furthermore for each link $l\in\mathbf{L}$ where $|\mathcal{G}^l|=2$ or $|\mathcal{G}^l|=3$ the network manager chooses at random a user $k_l\notin\mathcal{G}^l$ and assigns the subsidy $\mathcal{Q}^l$ to user k_l. Let $l_1,l_2,\ldots,l_r$ be the set of links such that $|\mathcal{G}^{l_i}|=2$ or 3, $i=1,2,\ldots,r$, and let k_{l_i}, $i=1,2,\ldots,r$, be the corresponding users that receive $\mathcal{Q}^{l_i}$.

Based on the above, the tax (subsidy) paid (received) by user j, $j\in\mathcal{N}$, is the following. If $j\neq k_{l_1},k_{l_2},\cdots k_{l_r}$ then

$$t_j=\sum_{l\in\mathcal{R}_j}t_j^l, \tag{3.17}$$

where for each $l\in\mathcal{R}_j$, t_j^l is determined according to the cardinality of $\mathcal{G}^l$. If $j=k_{l_i}$, $i=1,2,\ldots,r$, then

$$t_{k_{l_i}}=\sum_{l\in\mathcal{R}_{k_{l_i}}}t_{k_{l_i}}^l+\mathcal{Q}^{l_i}. \tag{3.18}$$

where $\mathcal{Q}^{l_i}$ is defined by (3.15) and (3.16).

Note that $\mathcal{Q}^{l_i}$ is not controlled by user k_{l_i}, that is, $\mathcal{Q}^{l_i}$ does not depend on user k_{l_i}'s message/strategy. Thus, the presence (or absence) of $\mathcal{Q}^{l_i}$ does not influence the strategic behavior of user k_{l_i}. We have assumed here that the users $k_{l_1}, k_{l_2}, \ldots, k_{l_r}$, are distinct. Expressions similar to the above hold when the users $k_{l_1}, k_{l_2}, \ldots, k_{l_r}$ are not distinct.

Remark For each link $l \in \mathcal{L}$ with $|\mathcal{G}^l| = 2$ or 3 the network manager could equally divide the subsidy $\mathcal{Q}^l$ among all users not in $\mathcal{G}^l$ instead of randomly choosing one user $k \notin \mathcal{G}^l$. Any other division of the subsidy $\mathcal{Q}^l$ among users not in $\mathcal{G}^l$ would also work.

3.3.2 Discussion/Interpretation of the Mechanism

As pointed out in Sect. 3.2.2, the design of a decentralized resource allocation mechanism has to achieve the following goals. (1) It must induce strategic users to voluntarily participate in the allocation process. (2) It must induce strategic users to follow its operational rules. (3) It must result in optimal allocations at all equilibria of the induced game. (4) It must result in a balanced budget at all equilibria and off equilibrium.

Since the designer of the mechanism can not alter the users' utility functions, $u_i, i \in \mathcal{N}$, the only way it can achieve the aforementioned objectives is through the use of appropriate *tax incentives/tax functions*. At each link l, the tax incentive of our mechanism for user i consists of three components $\Delta_1^l(i)$, $\Delta_2^l(i)$ and $\Delta_3^l(i)$. We specify and interpret these components for *Case 3* (Eq. (3.13)). Similar interpretations hold for *Case 1* and *Case 2*.

For *Case 3* we have,

$$t_i^l := \Delta_1^l(i) + \Delta_2^l(i) + \Delta_3^l(i) \tag{3.19}$$

where

$$\Delta_1^l(i) := P_{-i}^l x_i \tag{3.20}$$

$$\Delta_2^l(i) := (p_i^l - P_{-i}^l)^2 - 2P_{-i}^l(p_i^l - P_{-i}^l)\left(\frac{\mathcal{E}_{-i}^l + x_i}{\gamma}\right) + \frac{1\{x_i > 0\}1\{\mathcal{E}_{-i}^l + x_i > 0\}}{1 - 1\{x_i > 0\}1\{\mathcal{E}_{-i}^l + x_i > 0\}} \tag{3.21}$$

$$\Delta_3^l(i) := \Phi_i^l \tag{3.22}$$

- $\Delta_1^l(i)$ specifies the amount user i has to pay for the bandwidth it gets at link l. It is important to note that the price per unit of bandwidth that a user pays is determined by the message/proposal of the other users using the same link. Thus, a user does not control the price it pays per unit of the service it receives.

- $\Delta_2^l(i)$ provides the following incentives to the users of a link: (1) To bid/propose the same price per unit of bandwidth at that link (2) To collectively request a total bandwidth that does not exceed the capacity of the link. The incentive provided to all users to bid the same price per unit of bandwidth is described by the term $(p_i^l - P_{-i}^l)^2$. The incentive provided to all users to collectively request a total bandwidth that does not exceed the link's capacity is captured by the term

$$\frac{1\{x_i > 0\}1\{\mathcal{E}_{-i}^l + x_i > 0\}}{1 - 1\{x_i > 0\}1\{\mathcal{E}_{-i}^l + x_i > 0\}}. \tag{3.23}$$

Note that a user is very heavily penalized if it requests a nonzero bandwidth, and, collectively, all the users of the link request a total bandwidth that exceeds the link's capacity. A joint incentive provided to all users to bid the same price per unit of bandwidth and to utilize the total capacity of the link is captured by the term

$$2P_{-i}(p_i^l - P_{-i}^l)\left(\frac{\mathcal{E}_{-i}^l + x_i}{\gamma}\right) \tag{3.24}$$

- $\Delta_3^l(i)$, The goal of this component is to lead to a balanced budget. That is,

$$\sum_{i \in \mathcal{G}^l} \left[\Delta_1^l(i) + \Delta_2^l(i)\right] \neq 0, \tag{3.25}$$

but,

$$\sum_{i \in \mathcal{G}^l} \left[\Delta_1^l(i) + \Delta_2^l(i) + \Delta_3^l(i)\right] = 0. \tag{3.26}$$

Note that, $\Delta_3^l(i)$ is not controlled by user i's messages (simply because there is no term in $\Delta_3^l(i)$ under the control of user i), so $\Delta_3^l(i)$ does not have any influence on the strategic behavior of the user.

As indicated in (3.26), when the number of users at link $l \in \mathbf{L}$ is larger than three, i.e. $|\mathcal{G}^l| > 3$, the mechanism is budget-balanced at that link, that is $\sum_{i \in \mathcal{G}^l} t_i^l = 0$. When $|\mathcal{G}^l| = 2, 3$ the mechanism is not budget balanced at link l. The amount $\mathcal{Q}^l = -\sum_{\substack{i \in \mathcal{G}^l \\ |\mathcal{G}^l|=2,3}} t_i^l$, is given as subsidy to a randomly chosen user, say j, that does not compete for resources at link l. Such money transfers results in an overall balanced budget, and are always possible whenever $N > 3$. Furthermore, the money transfered to user j does not alter j's strategic behavior since $\mathcal{Q}^l$ does not depend on user j's strategy. The existence of the term $\mathcal{Q}_{l_j}$ in the tax function couples the games that are taking place at various links of the network. The presence of $\mathcal{Q}^{l_j}$ implies that the designer of the mechanism must not consider links individually; for the allocation of resources at certain links (specially those links l with $|\mathcal{G}^l| = 2, 3$)

the designer must consider network users that do not compete for resources in those links.

3.4 Properties of the Mechanism

We prove that the mechanism proposed in Sect. 3.3 has the following properties: (P1) It implements the solution of Problem **Max** in Nash equilibria. (P2) It is individually rational. (P3) It is budget-balanced at every feasible allocation, that is the mechanism is budget-balanced at allocations corresponding to all NE messages as well as those corresponding to off-equilibrium messages. We also prove the existence of NE of the game induced by the mechanism and characterize all of them.

We establish the above properties by proceeding as follows. First we prove that all Nash equilibria of the game induced by the game form/mechanism of Sect. 3.3 result in feasible solutions of the centralized problem **Max**, (Theorem 3.1). Then, we show that network users voluntarily participate in the allocation process. We do this by showing that the allocations they receive at all Nash equilibria of the game induced by the game form of Sect. 3.3 are weakly preferred to the (0, 0) allocation they receive when they do not participate in the allocation process (Theorem 3.4). Afterwards, we establish that the mechanism is budget-balanced at all Nash equilibria; we also prove that the mechanism is budget-balanced off equilibrium (Lemma 3.2). Finally, we show that the mechanism implements in Nash equilibria the solution of the centralized allocation problem **Max** (Theorem 3.5).

We present the proofs of the following theorems and lemmas in Appendix A.

Theorem 3.1 (FEASIBILITY): *If $m^* = (\boldsymbol{x}^*, \boldsymbol{p}^*)$ is a NE point of the game induced by the game form and the users' utility (outcome) functions presented in Sect. 3.3, then the allocation $\boldsymbol{x}^*$ is a feasible solution of Problem **Max**.*

The following lemma presents some key properties of NE prices and rates.

Lemma 3.2 *Let $m^* = (\boldsymbol{x}^*, \boldsymbol{p}^*)$ be a NE. Then for every $l \in \mathbf{L}$ and $i \in \mathcal{G}^l$, we have,*

$$p_i^{*l} = p_j^{*l} = P_{-i}^{*l} := p^{*l}, \tag{3.27}$$

$$p^{*l}\left(\frac{\mathcal{E}^{*l}}{\gamma}\right) = 0, \tag{3.28}$$

$$\left.\frac{\partial t_i^l}{\partial x_i}\right|_{m=m^*} = p^{*l}, \tag{3.29}$$

where $\mathcal{E}^{*l} = \sum_{i \in \mathcal{G}^l} x_i^* - c^l$.

An immediate consequence of Lemma 3.2 is the following. At every NE point m^* of the game induced by the mechanism the tax function has the following form,

$$t_i^l(m^*) = \begin{cases} p^{*l}x_i^* & \text{if } |\mathcal{G}^l| = 2; \\ p^{*l}(x_i^* - x_{-i}^*) + \frac{(p^{*l})^2(c_l - \mathcal{E}_{-i}^{*l})}{\gamma} & \text{if } |\mathcal{G}^l| = 3; \\ p^{*l}(x_i^* - x_{-i}^*) & \text{if } |\mathcal{G}^l| > 3. \end{cases} \tag{3.30}$$

Thus, by (3.15–3.18) and Lemma 3.2 we have,

$$t_i(m^*) = \sum_{l \in \mathcal{R}_i} t_i^l(m^*), \tag{3.31}$$

for $i \neq k_{l_1}, k_{l_2}, \dots, k_{l_r}$, (cf Sect. 3.3), and for $i = k_{l_j}$, $j = 1, 2, \dots, r$,

$$t_{k_{l_j}}(m^*) = \mathcal{Q}^{*l_j} + \sum_{l \in \mathcal{R}_{k_{l_j}}} t_{k_{l_j}}^l(m^*). \tag{3.32}$$

In the following lemma, we prove that the proposed mechanism is always budget balanced.

Lemma 3.3 *The proposed mechanism/game form is always budget balanced at every feasible allocation. That is, the mechanism is budget-balanced at all allocations corresponding to NE messages as well as at messages that are off equilibrium.*

The next result asserts that the mechanism/game form proposed in Sect. 3.3 is individually rational.

Theorem 3.4 (INDIVIDUAL RATIONALITY)*: The game form specified in Sect. 3.3 is individually rational, that is at every NE of the game induced by the mechanism the corresponding allocation* $(\boldsymbol{x}^*, \mathbf{t}^*)$ *is weakly preferred by all users to the initial allocation* $(0, 0)$.

Finally, we prove that the mechanism of Sect. 3.3 implements in NE the correspondence π defined by the solution of Problem **Max**.

Theorem 3.5 (NASH IMPLEMENTATION)*: Consider any NE* m^* *of the game induced by the mechanism of Sect. 3.3. Then, the allocation* $(\boldsymbol{x}^*, \mathbf{t}^*)$ *corresponding to* m^* *is an optimal solution of the centralized problem* ***Max***.

EXISTENCE AND CHARACTERIZATION OF THE NASH EQUILIBRIA:
So far, we have assumed the existence of NE of the game induced by the proposed game form/mechanism. In the following theorem, we prove that NE exist and characterize all of them.

Theorem 3.6 *Let $(x_1^*, x_2^*, \cdots, x_N^*)$ be an optimal solution of Problem* **Max** *and λ^{*l}, $l \in \mathbf{L}$, be the corresponding Lagrange multipliers of the Karush-Kuhn-Tucker (KKT) conditions. Then*

$$m^* := (x_1^*, x_2^*, \cdots, x_N^*, p^{*l_1}, p^{*l_2}, \cdots, p^{*l_L})$$

with $p^{*l} = \lambda^{*l}, l \in \mathbf{L}$ is a NE of the game induced by the proposed game form.

3.5 Implementation of the Decentralized Mechanism

We present one way of implementing the proposed mechanism at equilibrium. Consider an arbitrary link l of the network. The users of that link communicate their equilibrium messages to one another and to the network manager. The network manager determines the rate and tax (or subsidy) of each user and announces this information to the user. The users $i, i \in \mathcal{N}$ with tax $t_i^l > 0$ pay the amount t_i^l to the network manager; the network manager redistributes the amount of money it receives to the users $j \in \mathcal{N}$ with $t_j^l < 0$. In the situation where the number of users in the link is equal to two (resp. three) the network manager chooses randomly a user not using that link to whom it gives the subsidy $Q^{*\{l:|\mathcal{G}^l|=2\}}$ (resp. $Q^{*\{l:|\mathcal{G}^l|=3\}}$) defined by (3.15) (resp. (3.16)). The above described process is repeated/takes place at every network link. This process implements the mechanism described in the chapter at equilibrium.

3.6 An Extension

So far we required that the users' utility functions to be concave. We now weaken this requirement; we assume that the users' utilities are quasi-concave. We consider the game form proposed in Sect. 3.3. By repeating the arguments of Theorem 3.1, Lemma 3.2, Lemma 3.3 and Theorem 3.4 we can show that: every NE of the game induced by the game form is feasible; the game form/mechanism is individually rational and budget-balanced at all feasible allocations, i.e. at every NE and off equilibrium. In the following theorem we prove that every NE of the game induced by the proposed game form results in a Walrasian Equilibrium (WE), [1].

Theorem 3.7 *Consider the game $(\mathcal{M}, f, \mathcal{V}_i, i = 1, 2, \ldots, N)$, induced by the game form of Sect. 3.3, with continuous and quasi-concave utilities u_i, $i \in \mathcal{N}$. Then, every NE, m^*, of this game results in a Walrasian equilibrium, hence a Pareto optimal allocation (x^*, t^*).*

References

1. Mas-Colell A, Whinston MD, Green JR (2005) Microeconomic theory. Oxford University Press
2. Acemoglu D, Bimpkis K, Ozdaglar A (2009) Price and capacity competition. Game Econ Behav 50:1–26
3. Acemoglu D, Ozdaglar A (2007) Competition and efficiency in congested markets. Math Oper Res 32:1–31
4. Acemoglu D, Johari R, Ozdaglar A (2007) Partially optimal routing. IEEE J Sel Areas Commun 25(6):1148–1160
5. Kelly F (1994) On tariffs, policing and admission control for multi-service networks. Oper Res Lett 15:1–9
6. Kelly F, Maulloo A, Tan D (1998) Rate control for communication networks: shadow prices, proportional fairness and stability. Oper Res Soc 49:237–252
7. Hajek B, Yang S (2004) Strategic buyers in a sum-bid game for flat networks, preprint
8. Johari R, Tsitsiklis J (2004) Efficiency loss in a network resource allocation game. Math Oper Res 29(3):407–435
9. Johari R, Tsitsiklis J (2005) Communication requirement of vcg-like mechanisms in convex environments. In: Proceedings of the 43rd annual allerton conference on communication, control and computing
10. Lazar A, Semret N (1997) The progressive second price auction mechanism for network resource sharing. In: Proceedings of the international symposium on dynamic games and applications
11. Lazar A, Semret N (1999) Design and analysis of the progressive second price auction for network bandwidth sharing, Telecommunication systems—special issue on network economics
12. Maheswaren R, Basar T (2003) Nash equilibrium and decentralized negotiation in auctioning divisible resources. J Group Decis Negot 13(2)
13. Maheswaren R, Basar T (2004) Social welfare of selfish agents: Motivating efficiency for divisible resources. In: Proceedings of control and decision conferences (CDC)
14. Stoenescu T, Ledyard J (2008) Nash implementation for resource allocation network problems with production
15. Yang S, Hajek B (2007) Vcg-kelly mechanisms for allocation of divisible goods: adapting vcg mechanisms to one-dimensional signals. IEEE J Sel Areas Commun 25(6):1237–1243
16. Jain R, Walrand J (2010) An efficient nash-implementation mechanism for divisible resource allocation. Automatica 46(8):1276–1283
17. Vickrey W (1961) Counter speculation, auctions, and sealed tenders. J Finance 16:8–37
18. Clarke E (1971) Multipart pricing of public good. Public Choice 2:19–33
19. Groves T (1973) Incentive in teams. Econometrica 41(4):617–631
20. Yang S, Hajek B (2006) Revenue and stability of a mechanism for efficient allocation of a divisible good

Chapter 4
Power Allocation and Spectrum Sharing in Multi-User, Multi-Channel Systems

4.1 Introduction

As wireless communication devices become more pervasive, the demand for the frequency spectrum that serves as the underlying medium grows. Traditionally, the problem of allocating the resource of the frequency spectrum has been handled by granting organizations and companies licenses to broadcast at certain frequencies. This rigid approach leads to significant under-utilization of this scarce resource. Moreover, frequency utilization varies significantly with time and location. A cognitive radio is a wireless communication device that is aware of its capabilities, environment, and intended use, and can also learn new waveforms, models, or operational scenarios [1]. Recently, the Federal Communications Commission (FCC) has established rules (see [2]) that describes how cognitive radios can lead to more efficient use of the frequency spectrum. These rules along with the cognitive radio's features and the fact that information in the wireless network is decentralized and users may be strategic give rise to a wealth of important and challenging research issues associated with power allocation and spectrum sharing.

In this chapter we investigate a power allocation and spectrum sharing problem arising in multi-user, multi-channel systems with decentralized information and strategic users. The key issues and challenges associated with this problem have been discussed in Sect. 1.2 of the thesis. Here we formulate the problem as a public goods allocation with strategic users. We propose a game form/mechanism for the solution of the problem, we analyze the mechanism's properties and compare our results to the existing literature on power allocation and spectrum sharing problem with strategic users.

A. Kakhbod, *Resource Allocation in Decentralized Systems with Strategic Agents*, Springer Theses, DOI: 10.1007/978-1-4614-6319-1_4,
© Springer Science+Business Media New York 2013

4.1.1 Contribution of the Chapter

The main contribution of this chapter in power allocation and spectrum sharing is the discovery of a mechanism/game form which possesses the following properties.

(P1) The allocation corresponding to every NE of the game induced by the game form/mechanism results in a Lindahl equilibrium, that is, it is Pareto optimal. Conversely, every Lindahl equilibrium results in a NE of the game induced by the proposed game form/mechanism.
(P2) It is individually rational, i.e., every user participates voluntarily in the game induced by the mechanism.
(P3) It is budget balanced at every NE of the game induced by it as well as at all off equilibrium messages that result in feasible allocations.

All the above desirable properties are achieved without any assumption about, concavity, monotonicity or quasi-linearity of the users' utility functions.

4.1.2 Comparison with Related Work

The results presented in this chapter are distinctly different from those currently existing in the literature for the reasons we explain below.

Most of previous work within the context of competitive power allocation games has investigated Gaussian interference games [3, 4], that is, situations where the users operate in a Gaussian noise environment. In a Gaussian interference game, every user can spread a fixed amount of power arbitrarily across a continuous bandwidth, and attempts to maximize its total rate over all possible power allocation strategies. In [4], the authors proved the existence and uniqueness of a NE for a two-player version of the game, and provided an iterative water-filling algorithm to obtain the NE. This work was extended in [3], where it was shown that the aforementioned pure NE can be quite inefficient, but by playing an infinitely repeated game system performance can be improved. Our results are different from those in [3, 4] because: (i) The users are allowed to transmit at a discrete set of frequencies, and the power allocated at each frequency most be chosen from a discrete set. (ii) The unique pure NE of the one-stage game in [3, 4] does not necessarily result in a Pareto optimal allocation. (iii) Most of the NE of the repeated game in [3] result in allocations that are not Pareto optimal.

In [5], the authors presented a market-based model for situations where every user can only use one or more than one frequency bands, and the game induced by their proposed game form is super modular. They developed/presented a distributed best response algorithm that converges to a NE. However, in general the Nash equilibria of the game induced by their mechanism are not efficient, that is, they do not always result in optimal centralized power allocations, or Pareto optimal allocations.

In [6] the authors investigated the case where all users have the same utility function and each user can only use one frequency band. They proved the existence of a NE in the game resulting from the above assumptions. The NE is, in general, not efficient. The results in [6] critically depend on the fact that the users' utilities are identical and monotonic; these constraints are not present in our model.

The game form/mechanism we have proposed/analyzed in this chapter is in the category of the mechanisms that economists created for public good problems [7–9], but it is distinctly different form all of them because the allocation spaces in our formulation are discrete. To the best of our knowledge, the game form we presented in this chapter is the first mechanism for power allocation and spectrum sharing in multi-user, multi-channel systems with strategic users that achieves all three desirable properties (P1)–(P3). Furthermore, we do not impose any assumption about, concavity, differentiability, monotonicity or quasi-linearity of the users' utility functions.

4.1.3 Organization of the Chapter

The rest of the chapter is organized as follows. In Sect. 4.2 we present our model, describe the assumptions on the model's information structure and state our objective. In Sect. 4.3 we describe the allocation game form/mechanism we propose for the solution of our problem. In Sect. 4.4 we interpret the components of the proposed game form/mechanism. In Sect. 4.5 we investigate the properties of the proposed game form.

4.2 The Model and Objective

4.2.1 The Model

We consider N users/agents communicating over f frequency bands. Let $\mathcal{N} := \{1, \dots, N\}$ be the set of users, and $\mathbf{F} := \{1, 2, \dots, f\}$ the set of frequency bands. Each user $i, i \in \mathcal{N}$, is a communicating pair consisting of one transmitter and one receiver. There is one additional agent, the $(N+1)$th agent, who is different from all the other N agents/users and whose role will be described below. Each user has a fixed total power $\bar{W}$ which he can allocate over the set $\mathbf{F}$ of frequency bands. Let $p_i^j, i \in \mathcal{N}, j \in \mathbf{F}$ denote the power user i allocates to frequency band j. The power $p_i^j, i \in \mathcal{N}, j \in \mathbf{F}$ must be chosen from the set $\mathbf{Q} := \{\emptyset, Q_1, Q_2, \dots, Q_l\}$ where $Q_k > 0, 1 \leq k \leq l$, and $\emptyset$ means that user i does not use frequency band $j \in \mathbf{F}$ to communicate information. In other words, $\mathbf{Q}$ is a set of quantization levels that a user can use when he allocates power in a certain frequency band. Let $\bar{p}_i := (p_i^1, p_i^2, \dots, p_i^f), i \in \mathcal{N}$, denote a feasible bundle of power user i allocates over the frequency bands in $\mathbf{F}$. That is, $p_i^j \in \mathbf{Q}, \ \forall j \in \mathbf{F}$, and $\sum_{j \in \mathbf{F}} p_i^j \leq \bar{W}$. Let

$\mathrm{P} := (\bar{p}_1, \bar{p}_2, \ldots, \bar{p}_N)$ be a profile of feasible bundles of powers allocated by the N users over the frequency bands in $\mathbf{F}$; let Π denote the set of all feasible profiles P. Since the sets, $\mathcal{N}$, $\mathbf{F}$ and $\mathbf{Q}$ are finite, Π is finite. Let $|\Pi| = G_N$; we represent every feasible power profile by a number between 1 and G_N. Thus, $\Pi = \{1, 2, \ldots, G_N\}$. If user i allocates positive power in frequency band j, he may experience interference from those users who also allocate positive power in that frequency band. The intensity of the interference experienced by user $i, i \in \mathcal{N}$, depends on the power profiles used by the other users and the 'channel gains' h_{ji} between the other users $j, j \neq i$, and i. The satisfaction that user $i, i \in \mathcal{N}$, obtains during the communication process depends on his transmission power and the intensity of the interference he experiences. Consequently, user i's, $i \in \mathcal{N}$, satisfaction depends on the whole feasible bundle $k, k \in \Pi$, of power and is described by his utility function $V_i(k, t_i), i \in \mathcal{N}$, where $t_i \in \mathbb{R}$ represents the tax (subsidy) user i pays (receives) for communicating. One example of such a utility function is presented in the discussion following the assumptions. All taxes are paid to the $(N+1)$th agent who is not a profit maker; this agent acts like an accountant, collects the money from all users who pay taxes and redistributes it to all users who receive subsidies.

We now state our assumptions about the model, the users' utility functions, and the nature of the problem we investigate. Some of these assumptions are restrictions we impose, some others are a consequence of the nature of the problem we investigate. We comment on each of the assumptions we make after we state all of them.

(A1) We consider a static power allocation and spectrum sharing problem.

(A2) Each agent/user is aware of all the other users present in the system. Users talk to each other and exchange messages in a broadcast setting. That is, each user hears every other user's message; the $(N+1)$th agent hears all the other users' messages. After the message exchange process ends/converges, decision about power allocations at various frequency bands are made.

(A3) Each user's transmission at a particular frequency band creates interference to every user transmitting in the same frequency band.

(A4) The channel gains $h_{ji}(\hat{f}), j, i \in \mathcal{N}, \hat{f} \in \mathbf{F}$ are known to user $i, i \in \mathcal{N}$. The gains $h_{ji}(\hat{f}), j, i \in \mathcal{N}, \hat{f} \in \mathbf{F}$, do not change during the communication process.

(A5) Each user's utility $V_i(x, t_i), x \in \Pi \cup \{0\}$,[1] is decreasing in $t_i, t_i \in \mathbb{R}, i \in \mathcal{N}$. Furthermore, $V_i(x, t_i) \geq V_i(0, t_i)$ for any $t_i \in \mathbb{R}$ and $x \in \Pi$.

(A6) The utility function $V_i, i \in \mathcal{N}$, is user i's private information.

(A7) The quantization set $\mathbf{Q}$ is selected from $\mathcal{Q}$; the parameter $\bar{W}$ is selected from $\mathcal{W}$, and V_i is selected from $\mathcal{V}$ for all $i, i \in \mathcal{N}$. $\mathbf{Q}$, $\mathcal{Q}$, $\bar{W}$, $\mathcal{W}$ and $\mathcal{V}$ are common knowledge among all users.

(A8) Each user behaves strategically, that is, each user is selfish and attempts to maximize his own utility function under the constraints on the total power available to him, and on the set $\mathbf{Q}$ of quantization levels.

(A9) The representation/association of every feasible power profile by a number in the set Π is common knowledge among all users.

[1] The number zero denotes every non-feasible allocation.

We now briefly discuss each of the above assumptions. We restrict attention to the static power allocation and spectrum sharing problem (A1). The dynamic problem is a major open problem that we intend to address in the future. We assume that all users are in a relatively small area, so they can hear each other, are aware of the presence of one another, interfere with one another and exchange messages in a broadcast setting ((A2),(A3)). Since each user's satisfaction depends on his transmission power and the interference he experiences, his utility will depend on the whole power profile $x \in \Pi$; furthermore, the higher the tax a user pays, the lower is his satisfaction; moreover any feasible power allocation x, (i.e. $x \in \Pi$) is preferred to any non-feasible power allocation denoted by 0. All these considerations justify (A5). An example of $V_i(x, t_i)$ is

$$U_i\left(\frac{h_{ii}(1)p_i^1}{\frac{\mathcal{N}_0}{2}+\sum_{j,j\neq i}h_{ji}(1)p_j^1},\ldots,\frac{h_{ii}(f)p_i^f}{\frac{\mathcal{N}_0}{2}+\sum_{j,j\neq i}h_{ji}(f)p_j^f}\right)-t_i,$$

where $\frac{h_{ii}(k)p_i^k}{\frac{\mathcal{N}_0}{2}+\sum_{j,j\neq i}h_{ji}(k)p_j^k}$ is the Signal to Interference Ratio (SIR) in frequency band k. This example illustrates the following: (1) A user's utility function may explicitly depend on the channel gains h_{ji}, $j, i \in \mathcal{N}$; (2) User $i, i \in \mathcal{N}$, must know h_{ji}, $j \in \mathcal{N}$, so that he can be able to evaluate the impact of any feasible power profile $x \in \Pi$ that he proposes on his own utility. Thus, we assume that the channel gains h_{ji}, $j \in \mathcal{N}$, are known to user i, and this is true for every user i (A4). These channel gains have to be measured before the communication process starts. In the situation where users are cooperative h_{ji} can easily be determined; user j sends a pilot signal of a fixed power to user i, user i measures the received power and determines h_{ji}. When users are strategic/selfish, the measurement of h_{ji} can not be achieved according to the process described above, because user j may have an incentive to use a pilot signal other than the one agreed beforehand so that he can obtain an advantage over user i. In this situation procedures similar to ones described in [3] (Sect. 5) can be used to measure h_{ji}; we present a method, different from those proposed in [3], for measuring $h_{ji}(\hat{f})$, $j, i \in \mathcal{N}$, $\hat{f} \in \mathbf{F}$, after we discuss all the assumptions. In (A4) we further assume that h_{ji} do not change during the communication process. Such an assumption is reasonable when the mobile users move slowly and the variation of the channel is considerably slower than the duration of the communication process. Assumption (A8) is a behavioral one not a restriction on the model. Since according to (A8) users are strategic, each user may not want to reveal his own preference over the set of feasible power allocations, thus assumption (A6) is reasonable. It is also reasonable to assume that the function space where each user's utility comes from is the same for all users and common knowledge among all users (A7). The fact that a user's utility is his private information along with assumption (A8) have an immediate impact on the solution/equilibrium concepts that can be used in the game induced by any mechanism. We will address this issue when we define the objective of our problem. Assumption (A7) also ensures that each user uses the same quantization set. Furthermore, it states that each user knows the power available to every other user. The solution methodology presented in this chapter works also for the case where every user knows his total available power, has an upper bound on the power available to all other users, and this upper bound is common knowledge among all users. Assumption (A9) is necessary for the proposed game form/mechanism in this chapter; it ensures that each user interprets consistently the messages he receives from all other users.

In addition to the method described in [3], another method for determining the gains $h_{ji}(\hat{f})$, $j, i \in \mathcal{N}$, $\hat{f} \in \mathbf{F}$ is the following. We assume that the gain $h_{ij}(\hat{f})$ from the transmitter of pair i to the receiver of pair j is the same as $\bar{h}_{ji}(\hat{f})$, the gain from the receiver of pair j to the transmitter of pair i for all $i, j \in \mathcal{N}$ and $\hat{f} \in \mathbf{F}$. Before the power allocation and spectrum sharing process starts, the $(N+1)$th agent asks transmitter i and receiver j to communicate with one another at frequency $\hat{f}$ by using a fixed power $\bar{p}$, and to report to him their received powers. This communication process takes place as follows: First transmitter i sends a message with power $\bar{p}$ at frequency $\hat{f}$ to receiver j; then receiver j sends a message with power $\bar{p}$ at frequency $\hat{f}$ to transmitter i; finally transmitter i and receiver j report their received power to the $(N+1)$th agent. This process is sequentially repeated between transmitter i and receiver j for all frequencies $\hat{f} \in \mathbf{F}$. After transmitter i and receiver j complete the above-described communication process, the same process is repeated sequentially for all transmitter-receiver pairs (k, l), $k, l \in \mathcal{N}$, at all frequencies $\hat{f} \in \mathbf{F}$. The $(N+1)$th agents collects all the reports generated by the process described above. If the reports of any transmitter i and receiver $j (i \neq j,\ i, j \in \mathcal{N})$ differ at any frequency $\hat{f} \in \mathbf{F}$, then user i and user j are not allowed to participate in the power allocation and spectrum sharing process.

The above-described method for determining $h_{ji}(\hat{f})$, $i, j \in \mathcal{N}$, $\hat{f} \in \mathbf{F}$, provides an incentive to user i, $i \in \mathcal{N}$, to follow/obey its rules if user i does better by participating in the power allocation and spectrum sharing process than by not participating in it. Consequently, the method proposed for determining $h_{ji}(\hat{f})$, $i, j \in \mathcal{N}$, $\hat{f} \in \mathbf{F}$, will work if the game form we propose is individually rational. In this chapter we prove that individual rationality is one of the properties of the proposed game form.

4.2.2 Objective

The objective is to determine a game form/mechanism that, for any realization $(V_1, V_2, \ldots, V_N, \mathbf{Q}, \bar{W}) \in \mathcal{V}^N \times \mathcal{Q} \times \mathcal{W}$, possesses the following features:

(P1) It implements in NE the Pareto correspondence.

(P2) It is individually rational, i.e., the users voluntarily participate in the game induced by it.

(P3) It is budget balanced at all NE of the game it induces, as well as at all off equilibrium messages that result in feasible allocations.

We follow the philosophy of implementation theory (cf. Sect. 2.1.1) for the specification of our game form. We note (cf. Sect. 2.1.4) that the Pareto correspondence is implementable in NE. In the next section we present a game form/mechanism that achieves the above objective.

4.3 A Mechanism for Power Allocation and Spectrum Sharing

For the decentralized problem formulated in Sect. 4.2 we propose a game form the components of which are described as follows.

Message space $\mathcal{M} := \mathcal{M}_1 \times \mathcal{M}_2 \times \cdots \mathcal{M}_N$: The message/strategy space for user $i, i = 1, 2, \ldots, N$, is given by $\mathcal{M}_i \subseteq \mathbb{Z} \times \mathbb{R}_+$, where $\mathbb{Z}$ and $\mathbb{R}_+$ are the sets of integers and non-negative real numbers, respectively. Specifically, a message of user i is of the form, $m_i = (n_i, \pi_i)$ where $n_i \in \mathbb{Z}$ and $\pi_i \in \mathbb{R}_+$.

The meaning of the message space is the following. The component n_i represents the power profile proposed by user i; the component π_i denotes the price per unit of power user i is willing to pay per unit of the power profile n_i. The message n_i belongs to an extended set $\mathbb{Z}$ of power profiles. Every element/integer in $\mathbb{Z} - \Pi$ corresponds to a power profile that is non-feasible. Working with such an extended set of power profiles does not alter the solution of the original problem since, as we show in Sect. 4.5, all Nash equilibria of the game induced by the proposed mechanism correspond to feasible power allocations.

Outcome function $\hbar$: The outcome function $\hbar$ is given by, $\hbar : \mathcal{M} \to \mathbb{N} \times \mathbb{R}^N$. and is defined as follows. For any $m := (m_1, m_2, \ldots, m_N) \in \mathcal{M}$,

$$\hbar(m) = \hbar(m_1, m_2, \ldots, m_N)$$
$$= \left(\left[\mathbb{I}\left(\frac{\sum_{i=1}^N n_i}{N} \right) \right], t_1(m), \ldots, t_N(m) \right),$$

where $\mathbb{I}\left(\frac{\sum_{i=1}^N n_i}{N} \right)$ is the integer number closest (from above) to $\left(\frac{\sum_{i=1}^N n_i}{N} \right)$ and

$$\left[\mathbb{I}\left(\frac{\sum_{i=1}^N n_i}{N} \right) \right] = \begin{cases} \mathbb{I}\left(\frac{\sum_{i=1}^N n_i}{N} \right), & \text{if } \mathbb{I}\left(\frac{\sum_{i=1}^N n_i}{N} \right) \in \Pi; \\ 0, & \text{otherwise.} \end{cases}$$

The component $t_i, i = 1, 2, \ldots, N$, describes the tax (subsidy) that user i pays (receives). The tax(subsidy) for every user is defined as follows,

$$t_i(m) = \left\{ \mathbb{I}\left(\frac{\sum_{i=1}^N n_i}{N} \right) \left[\frac{\pi_{i+1} - \pi_{i+2}}{N} \right] + (n_i - n_{i+1})^2 \pi_i - (n_{i+1} - n_{i+2})^2 \pi_{i+1} \right\}$$

$$\times\, 1\left\{\mathbb{I}\left(\frac{\sum_{i=1}^{N} n_i}{N}\right) \in \Pi\right\} \tag{4.1}$$

where $1\{A\}$ denotes the indicator function of event A, that is, $1\{A\} = 1$ if A is true and $1\{A\} = 0$ otherwise, and $N + 1$ and $N + 2$ are to be interpreted as 1 and 2, respectively.

4.4 Interpretation of the Mechanism

As pointed out in Sect. 4.2, the design of an efficient resource allocation mechanism has to achieve the following goals. (i) It must induce strategic users to voluntarily participate in the allocation process. (ii) It must induce strategic users to follow its operational rules. (iii) It must result in Pareto optimal allocations at all equilibria of the induced game. (iv) It must result in a balanced budget at all equilibria and off equilibrium.
To achieve these goals we propose the tax incentive function described by Eq. (4.1). This function consists of three components, Ξ_1, Ξ_2 and Ξ_3, that is,

$$t_i(m) = \underbrace{\mathbb{I}\left(\frac{\sum_{i=1}^{N} n_i}{N}\right)\left[\frac{\pi_{i+1} - \pi_{i+2}}{N}\right]}_{\Xi_1} + \underbrace{(n_i - n_{i+1})^2 \pi_i}_{\Xi_2} \underbrace{-(n_{i+1} - n_{i+2})^2 \pi_{i+1}}_{\Xi_3} \tag{4.2}$$

The term Ξ_1 specifies the amount that each user must pay for the power profile which is determined by the mechanism. The price per unit of power, $\frac{\pi_{i+1} - \pi_{i+2}}{N}$, paid by user i, $i = 1, 2, \ldots, N$, is not controlled by that user. The terms Ξ_2 considered collectively provide an incentive to all users to propose the same power profile. The term Ξ_3 is not controlled by user i, its goal is to lead to a balanced budget.

4.5 Properties of the Mechanism

We prove the mechanism proposed in Sect. 4.3 has the properties (P1), (P2) and (P3) stated in Sect. 4.2.2 by proceeding as follows. First, we derive a property of every NE of the game induced by the mechanism proposed in Sect. 4.2, (Lemma 4.1); based on this result we determine the form of the tax (subsidy) at all Nash equilibria. Then, we show that every NE of the game induced by the proposed mechanism results in a feasible allocation, (Lemma 4.2). Afterward, we prove that the proposed mechanism is always budget balanced, (Lemma 4.3). Subsequently we show that users voluntarily participate in the game, by proving that the utility they receive at all NE is greater than or equal to zero, which is the utility they receive by not participating in the power allocation and spectrum sharing process, (Lemma 4.4). Finally, we show that every NE of the game induced by the mechanism proposed in Sect. 4.3 results in a Lindahl equilibrium ([10] Sect. 12.4.2); that is, every NE results in a Pareto optimal allocation (Theorem 4.5). Furthermore, we prove that every Lindahl equilibrium can be associated with a NE of the game induced by the proposed mechanism in Sect. 4.3, (Theorem 4.6).

We now proceed to prove the above-stated properties.

Lemma 4.1 *Let m^* be a NE of the game induced by the proposed mechanism. Then for every $i, i = 1, 2, \ldots, N$, we have*

$$(n_i^* - n_{i+1}^*)^2 \pi_i^* = 0. \tag{4.3}$$

Proof Since $m^* = ((n_1^*, \pi_1^*), (n_2^*, \pi_2^*), \ldots, (n_N^*, \pi_N^*))$ is a NE, the following holds for every $i, i = 1, 2, \ldots, N$, and $\forall\, m_i \in \mathcal{M}_i$,

$$V_i\left(\left[\mathbb{I}\left(\frac{\sum_{k=1}^N n_k^*}{N}\right)\right], t_i(m^*)\right) \geq V_i\left(\left[\mathbb{I}\left(\frac{\sum_{\substack{k=1\\k\neq i}}^N n_k^* + n_i}{N}\right)\right], t_i(m_i, m_{-i}^*)\right), \tag{4.4}$$

where $m_{-i} := (m_1, m_2, \ldots, m_{i-1}, m_{i+1}, \ldots, m_N)$.
Set n_i equal to n_i^*; then for every $\pi_i \geq 0$ Eq. (4.4) along with (4.1) imply

$$V_i\left(\left[\mathbb{I}\left(\frac{\sum_{k=1}^N n_k^*}{N}\right)\right], t_i(m^*)\right) \geq V_i\left(\left[\mathbb{I}\left(\frac{\sum_{k=1}^N n_k^*}{N}\right)\right], t_i(m_i, m_{-i}^*)\right) \tag{4.5}$$

where

$$t_i(m^*) = \left\{\mathbb{I}\left(\frac{\sum_{k=1}^N n_k^*}{N}\right)\left[\frac{\pi_{i+1}^* - \pi_{i+2}^*}{N}\right] + (n_i^* - n_{i+1}^*)^2 \pi_i^* - (n_{i+1}^* - n_{i+2}^*)^2 \pi_{i+1}^*\right\} \times 1\left\{\mathbb{I}\left(\frac{\sum_{i=1}^N n_i^*}{N}\right) \in \Pi\right\} \tag{4.6}$$

$$t_i((n_i^*, \pi_i), m_{-i}^*) = \left\{\mathbb{I}\left(\frac{\sum_{k=1}^N n_k^*}{N}\right)\left[\frac{\pi_{i+1}^* - \pi_{i+2}^*}{N}\right] + (n_i^* - n_{i+1}^*)^2 \pi_i - (n_{i+1}^* - n_{i+2}^*)^2 \pi_{i+1}^*\right\} \times 1\left\{\mathbb{I}\left(\frac{\sum_{i=1}^N n_i^*}{N}\right) \in \Pi\right\} \tag{4.7}$$

Since V_i is decreasing in t_i Eq. (4.5) along with (4.6) and (4.7) yield

$$\pi_i^*(n_i^* - n_{i+1}^*)^2 \leq \pi_i (n_i^* - n_{i+1}^*)^2 \quad \forall\, \pi_i \geq 0. \tag{4.8}$$

Therefore, $\pi_i^*(n_i^* - n_{i+1}^*)^2 = 0$ for every $i, i = 1, 2, \ldots, N$, at every NE m^*. □

An immediate consequence of Lemma 4.1 is the following. At every NE m^* of the game induced by the mechanism the tax function $t(m^*)$ has the form

$$t_i(m^*) = \mathbb{I}\left(\frac{\sum_{k=1}^N n_k^*}{N}\right)\left[\frac{\pi_{i+1}^* - \pi_{i+2}^*}{N}\right] \times 1\left\{\mathbb{I}\left(\frac{\sum_{i=1}^N n_i^*}{N}\right) \in \Pi\right\}. \tag{4.9}$$

In the following lemma, we show that every NE of the game induced by the proposed mechanism is feasible.

Lemma 4.2 *Every NE of the game induced by the proposed mechanism results in a feasible allocation.*

Proof We prove the assertion of the lemma by contradiction. Let m^* be a NE for the game induced by the mechanism. Suppose m^* does not result in a feasible allocation, i.e., $\mathbb{I}\left(\frac{\sum_{i=1}^N n_i^*}{N}\right) \notin \{1, 2, 3, \ldots, G_N\}$. Then $\left[\mathbb{I}\left(\frac{\sum_{j=1}^N n_j^*}{N}\right)\right] = 0$. Since $\sum_{j=1}^N (\pi_{j+1}^* - \pi_{j+2}^*) = 0$, there exists $i, i \in \{1, 2, \ldots, N\}$, such that

$$\pi_{i+1}^* - \pi_{i+2}^* \le 0. \tag{4.10}$$

Keep m_{-i}^* fixed and define $m_i = (n_i, \pi_i)$ as follows; set $\pi_i = 0$, and choose n_i such that $\mathbb{I}\left(\frac{\sum_{j=1, j\neq i}^N n_j^* + n_i}{N}\right) \in \Pi$. Now, Eq. (4.1) yield that

$$t_i(m_i, m_{-i}^*) \le 0. \tag{4.11}$$

Equation (4.11) along with Lemma 4.1 and assumption (A5) result in

$$\begin{aligned} V_i(0, 0) &= V_i\left(\left[\mathbb{I}\left(\frac{\sum_{j=1}^N n_j^*}{N}\right)\right], t_i(m^*)\right) \\ &< V_i\left(\left[\mathbb{I}\left(\frac{\sum_{j=1, j\neq i}^N n_j^* + n_i}{N}\right)\right], t_i(m_i, m_{-i}^*)\right). \end{aligned} \tag{4.12}$$

But (4.12) is in contradiction with the fact that m^* is a NE. Therefore, every NE of the game induced by the proposed mechanism results in a feasible allocation. □

In the following lemma, we show that the proposed mechanism is always budget balanced.

Lemma 4.3 *The proposed mechanism is always budget balanced.*

Proof To have a balanced budget it is necessary and sufficient to satisfy $\sum_{i=1}^N t_i(m_i) = 0$. It is easy to see that budget balance always holds since from

(4.1) we have

$$\sum_{i=1}^{N} t_i(m) = \sum_{i=1}^{N} \mathbb{I}\left(\frac{\sum_{i=1}^{N} n_i}{N}\right)\left[\frac{\pi_{i+1} - \pi_{i+2}}{N}\right] + \sum_{i=1}^{N} \left((n_i - n_{i+1})^2 \pi_i - (n_{i+1} - n_{i+2})^2 \pi_{i+1}\right) = 0. \tag{4.13}$$

The last equality in (4.13) holds, because

$$\sum_{i=1}^{N} (\pi_{i+1} - \pi_{i+2}) = \sum_{i=1}^{N} \left((n_i - n_{i+1})^2 \pi_i - (n_{i+1} - n_{i+2})^2 \pi_{i+1}\right) = 0.$$

□

The next result asserts that the mechanism/game form proposed in Sect. 4.3 is individually rational.

Lemma 4.4 *The game form specified in* Sect. 4.3 *is individually rational, i.e., at every NE m^* the corresponding allocation* $\left(\mathbb{I}\left(\frac{\sum_{i=1}^{N} n_i^*}{N}\right), t_1(m^*), t_2(m^*), \ldots, t_N(m^*)\right)$ *is weakly preferred by all users to their initial endowment* $(\emptyset, 0)$.

Proof We need to show that $V_i\left(\mathbb{I}\left(\frac{\sum_{i=1}^{N} n_i^*}{N}\right), t_i(m^*)\right) \geq V_i(\emptyset, 0) = 0$ for every $i, i = 1, 2, \ldots, N$. By the property of every NE, it follows that for every $i \in \mathcal{N}$ and $(n_i, \pi_i) \in \mathcal{M}_i$,

$$V_i\left(\mathbb{I}\left(\frac{\sum_{k=1}^{N} n_k^*}{N}\right), t_i(m^*)\right) \geq V_i\left(\left[\mathbb{I}\left(\frac{\sum_{\substack{k=1 \\ k \neq i}}^{N} n_k^* + n_i}{N}\right)\right], t_i((n_i, \pi_i), m_{-i}^*)\right). \tag{4.14}$$

Choosing n_i sufficiently large so that $\mathbb{I}\left(\frac{\sum_{\substack{k=1 \\ k \neq i}}^{N} n_k^* + n_i}{N}\right) \notin \{1, 2, \ldots, G_N\}$, gives

$$\left[\mathbb{I}\left(\frac{\sum_{\substack{k=1 \\ k \neq i}}^{N} n_k^* + \hat{n}_i}{N}\right)\right] = 0, \tag{4.15}$$

and

$$t_i((n_i, \pi_i), m^*_{-i}) = 0. \tag{4.16}$$

because of (4.1). Consequently, (4.15) and (4.16) establish that

$$V_i\left(\mathbb{I}\left(\frac{\sum_{k=1}^{N} n_k^*}{N}\right), t_i(m^*)\right) \geq V_i(\emptyset, 0) = 0. \tag{4.17}$$

□

In the following theorem, we show that every NE of the game induced by the mechanism proposed in Sect. 4.3 results in a Lindahl equilibrium.

Theorem 4.5 *Suppose that an allocation Ψ_m, for any $m \in \mathcal{M}$, is determined as follows*

$$\Psi_m := (\Lambda(m), t_1(m), \ldots, t_N(m), L_1, \ldots, L_N)$$

where $\Lambda(m) := \left[\mathbb{I}\left(\frac{\sum_{k=1}^{N} n_k}{N}\right)\right]$, *for each* $i, i = 1, 2, \ldots, N, t_i(m)$ *is defined by* (4.1), *and*

$$L_i := \frac{\pi_{i+1} - \pi_{i+2}}{N}. \tag{4.18}$$

Then Ψ_{m^*} is a Lindahl equilibrium corresponding to the NE

$$m^* = \left((n_1^*, \pi_1^*), (n_2^*, \pi_2^*), \ldots, (n_N^*, \pi_N^*)\right)$$

of the game induced by the proposed mechanism.

Proof Ψ_{m^*} defines a Lindahl equilibrium if it satisfies the following three conditions ([10] Sect. 12.4.2)

1. *(C1)*: $\sum_{i=1}^{N} L_i^* = 0$.
2. *(C2)*: $\sum_{i=1}^{N} t_i(m^*) = 0$.
3. *(C3)*: For all $i, i = 1, 2, \ldots, N$, $\left(\mathbb{I}\left(\frac{\sum_{k=1}^{N} n_k^*}{N}\right), t_i(m^*)\right)$ is a solution of the following optimization problem:

$$\begin{array}{ll} \max_{x,t_i} & V_i(x, t_i) \\ \text{subject to} & x\, L_i^* = t_i \\ & x \in \Pi. \end{array} \tag{4.19}$$

By simple algebra we can show that conditions 1 and 2 are satisfied. We need to prove that condition 3 is also satisfied. We do this by contradiction. Suppose $\left(\mathbb{I}\left(\frac{\sum_{k=1}^{N} n_k^*}{N}\right), t_i(m^*)\right)$ is not a solution of the optimization problem defined by (4.19) for all i. Then, for some user $i, i \in \{1, 2, \ldots, N\}$, there is a power profile

$\zeta \in \Pi$ and $\zeta \neq \mathbb{I}\left(\frac{\sum_{k=1}^{N} n_k^*}{N}\right)$ such that

$$V_i\left(\mathbb{I}\left(\frac{\sum_{k=1}^{N} n_k^*}{N}\right), \left[\mathbb{I}\left(\frac{\sum_{k=1}^{N} n_k^*}{N}\right)\right] L_i^*\right) < V_i(\zeta, \zeta\, L_i^*). \tag{4.20}$$

Now choose $\bar{\pi}_i = 0$ and $\bar{n}_i = \mathbb{I}\left(N\zeta - \sum_{\substack{j=1 \\ j\neq i}}^{N} n_j^*\right)$. Using Eqs. (4.1) and (4.3) together with the fact that $\bar{\pi}_i = 0$ we obtain

$$t_i((\bar{n}_i, \bar{\pi}_i), m_{-i}^*) = \zeta \left[\frac{\pi_{i+1}^* - \pi_{i+2}^*}{N}\right] = \zeta\, L_i^*. \tag{4.21}$$

Then, because of (4.20) and (4.21) we get

$$\begin{aligned} V_i(\zeta, \zeta\, L_i^*)Z &= V_i\left(\left[\mathbb{I}\left(\frac{\sum_{\substack{j=1 \\ j\neq i}}^{N} n_j^* + \bar{n}_i}{N}\right)\right], t_i((\bar{n}_i, \bar{\pi}_i), m_{-i}^*)\right) \\ &\geq V_i\left(\left[\mathbb{I}\left(\frac{\sum_{\substack{j=1 \\ j\neq i}}^{N} n_j^* + n_i^*}{N}\right)\right], t_i(m^*)\right) \end{aligned}$$

which is a contradiction, because

$$m^* = \left((n_1^*, \pi_1^*), (n_2^*, \pi_2^*), \ldots, (n_N^*, \pi_N^*)s\right)$$

is a NE of the game induced by the proposed game form. Consequently, $\left(\mathbb{I}\left(\frac{\sum_{k=1}^{N} n_k^*}{N}\right), t_i(m^*)\right)$ is a solution of the optimization problem defined by (4.19) for all i. Since Ψ_{m^*} satisfies *(C1)–(C3)* it defines a Lindahl equilibrium. The allocation

$$\left\{\mathbb{I}\left(\frac{\sum_{k=1}^{N} n_k^*}{N}\right), t_1(m^*), t_2(m^*), \ldots, t_N(m^*)\right\}$$

is also Pareto optimal ([10] Theorem (12.4.1)). □

Finally, we establish that any Lindahl equilibrium can be associated with a NE of the game induced by the proposed mechanism.

Theorem 4.6 *Let $\Psi = (\Lambda^\ell, t_1^\ell, t_2^\ell, \ldots, t_N^\ell, L_1^\ell, L_2^\ell, \ldots, L_N^\ell)$ be a Lindahl equilibrium. Then, there does exist a NE m^* of the game induced by the proposed mechanism so that*

$$\hbar(m^*) = \left(\Lambda^\ell, t_1^\ell, t_2^\ell, \ldots, t_N^\ell\right) \tag{4.22}$$

where for every $i, i = 1, 2, \ldots, N$, $t_i(m^*) = \Lambda^\ell \; L_i^\ell$.

Proof Consider the message profile m^* such that for every $i, i = 1, 2, \ldots, N, m_i^* = (n_i^*, \pi_i^*)$ and, $\forall i, i = 1, 2, \ldots, N, n_i^* = (\Lambda^\ell)$ and π_i^*'s are the solution of the following system of equations,

$$L_1^\ell = \frac{\pi_2^* - \pi_3^*}{N}, \quad L_2^\ell = \frac{\pi_3^* - \pi_4^*}{N}, \ldots, L_N^\ell = \frac{\pi_1^* - \pi_2^*}{N}. \tag{4.23}$$

Choosing π_1^* sufficiently large guarantees that the following is a feasible solution to (4.23), i.e., $\pi_i \geq 0, \forall\, i, \pi_1^* =$ sufficiently large, $\pi_2^* = \pi_1^* - L_N^\ell$ and

$$\pi_i^* = (i-1)\pi_1^* - \left(L_N^\ell + \sum_{j=1}^{i-2} L_j^\ell\right) \quad i, \; 3 \leq i \leq N.$$

Furthermore,

$$\Lambda^\ell = \left[\mathbb{I}\left(\frac{\sum_{k=1}^N n_k^*}{N}\right)\right]. \tag{4.24}$$

To complete the proof, we need to prove that m^* is a NE of the game induced by the mechanism. For that matter, it is enough to show that, for every $i, i = 1, 2, \ldots, N$,

$$V_i\left(\left[\mathbb{I}\left(\frac{\sum_{k=1}^N n_k^*}{N}\right)\right], t_i(m^*)\right) \geq V_i\left(\left[\mathbb{I}\left(\frac{\sum_{k=1, k\neq i}^N n_k^* + n_i}{N}\right)\right], t_i(m_{-i}^*, m_i)\right)$$
$$\forall\, m_i \in \mathcal{M}. \tag{4.25}$$

Equation (4.9) along with Eq. (4.23) imply $t_i(m^*) = L_i^\ell \left[\mathbb{I}\left(\frac{\sum_{k=1}^N n_k^*}{N}\right)\right]$. Furthermore, positivity of $(n_{i+1}^* - n_i)^2 \pi_i$ together with fact that V_i is decreasing in t_i give that

$$V_i\left(\xi, L_i^\ell \xi\right) \geq V_i\left(\xi, L_i^\ell \xi + (n_{i+1}^* - n_i)^2 \pi_i\right) \quad \forall \xi, \xi \in \Pi.$$

Moreover, since Ψ is a Lindahl equilibrium, *(C3)* implies that the following holds for every $i, i = 1, 2, \ldots, N$,

$$V_i\left(\Lambda^\ell, L_i^\ell \Lambda^\ell\right) \geq V_i\left(\xi, L_i^\ell \; \xi\right) \quad \forall \xi \in \Pi, \tag{4.26}$$

Consequently, the fact that $t_i(m^*) = L_i^\ell \left[\mathbb{I}\left(\frac{\sum_{k=1}^N n_k^*}{N} \right) \right]$ along with (4.25) and (4.26) result in

$$V_i\left(\left[\mathbb{I}\left(\frac{\sum_{k=1}^N n_k^*}{N}\right)\right], t_i(\mathbf{m}^*)\right) = V_i\left(\left[\mathbb{I}\left(\frac{\sum_{k=1}^N n_k^*}{N}\right)\right], L_i^\ell\left[\mathbb{I}\left(\frac{\sum_{k=1}^N n_k^*}{N}\right)\right]\right)$$
$$\geq V_i\left(\xi, L_i^\ell \xi\right)$$
$$\geq V_i\left(\xi, L_i^\ell \xi + (n_{i+1}^* - n_i)^2 \pi_i\right) \quad \forall \xi, \xi \in \Pi$$
$$= V_i\left(\left[\mathbb{I}\left(\frac{\sum_{j=1, j\neq i}^N n_j^* + n_i}{N}\right)\right], t_i(m_i, \mathbf{m}_{-i}^*)\right)$$
$$\forall m_i \in \mathcal{M}.$$

Therefore m^* is a NE of the game induced by the proposed mechanism. □

References

1. Neel JO, Reed JH, Gilles RP (2004) Convergence of cognitive radio networks. In: Proceedings of wireless communications and networking conference
2. FCC (2003) Facilitating opportunities for flexible, efficient, and reliableS spectrum use employing cognitive radio technologies, (ET docket no. 03–108). Technical report, FCC, Washington
3. Etkin R, Parekh A, Tse D (2007) Spectrum sharing for unlicensed bands. IEEE J Sel Areas Commun 25(3):517–528
4. Yu W, Ginis G, Cioffi JM (2002) Distributed multiuser power control for digital subscriber lines. IEEE J Sel Areas Commun 20(5):1105–1115
5. Huang J, Berry R, Honig ML (2006) Distributed interference compensation for wireless networks. IEEE J Sel Areas Commun 24(5):1074–1084
6. Tekin C, Liu M, Southwell R, Huang J, Ahmad S (2012) Atomic congestion games on graphs and their applications in networking. IEEE/ACM Trans Netw (to appear)
7. Walker M (1981) A simple incentive compatible scheme for attaining Lindahl allocations. Econometrica 49(1):65–71
8. Hurwicz L (1979) Outcome functions yielding Walrasian and Lindahl allocations at Nash equilibrium points. Rev Econ Stud 46:217–225
9. Stoenescu T, Ledyard J (2008) Nash implementation for resource allocation network problems with production, manuscript
10. Tian G (2010) Lecture notes microeconomic theory. On line

Chapter 5
Multi-Rate Multicast Service Provisioning

5.1 Introduction

5.1.1 Motivation and Challenges

Multicasting provides an efficient method of transmitting data in real time applications from one source to many users. The source sends one copy of a message to its users and this copy is replicated only at the branching points of a multicast tree. Real life examples of such multicast applications are audio/video broadcasting, teleconferencing, distributed databases, financial information, electronic newspapers, weather maps and experimental data. Conventional multicast studies the problem in which the rate received by all the users of the same multicast group is constant. The inherent problem with such a formulation is that a constant rate will overwhelm the slow receivers while starving the fast ones. Multi-rate multicast transmissions can be used to address this problem by allowing a receiver to obtain data at a rate that satisfies its requirements.

In this chapter we investigate the multi-rate multicast service provisioning problem in wired networks with arbitrary topology and strategic users. We formulate the problem as the combination of a market and a public goods allocations with strategic users. All existing literature on multi-rate multicast assumes non-strategic users. As we explain in Sect. 5.1.2 below, the nature of the problem suggests that strategic behavior may be beneficial to the users. Strategic behavior results in new challenges (conceptual and technical) in multi-rate multicast. The key issues and challenges associated with this problem have been discussed in Sect. 1.2 of the thesis. Here we propose a game form/mechanism for the solution of the problem, and analyze the mechanism's properties.

A. Kakhbod, *Resource Allocation in Decentralized Systems with Strategic Agents*, Springer Theses, DOI: 10.1007/978-1-4614-6319-1_5,
© Springer Science+Business Media New York 2013

5.1.2 Why is Strategic Behavior Justified?

Strategic behavior in multi-rate multicast can be justified as follows. The literature on multi-rate multicast with non-strategic users reveals that the problem has characteristics of the *free-rider* problem. That is, at any network link, a member of a multicast group is charged only if it requests the maximum rate/bandwidth within the group at that link. As a result of this feature of the problem, users are incentivized to misrepresent their demand for bandwidth; by slightly reducing its demand, a user can increase its overall utility because it slightly reduces its own satisfaction from the quality of service it receives, but pays considerably less tax. Thus, strategic behavior may result in higher overall utility for a user than non-strategic behavior.

5.1.3 Contribution of the Chapter

The main contributions of this chapter are:

1. The formulation of the multi-rate multicast service provisioning problem in wired networks with arbitrary topology and strategic users.
2. The discovery of a decentralized rate allocation mechanism for multi-rate multicast service provisioning in networks with arbitrary/general topology and strategic users, which possesses the following properties.

   (P1) It implements weakly the solution of the centralized multi-rate multicast service provisioning problem in Nash equilibria. That is, the allocation corresponding to each NE of the game induced by it, is a globally optimal solution of the corresponding centralized multi-rate multicast service provisioning problem.

   (P2) It is individually rational, that is, the network users/users voluntarily participate in the rate allocation process.

   (P3) It is budget balanced at all feasible allocations, that is, at all the allocations that correspond to NE messages/strategies as well as at all the feasible allocations that correspond to off-equilibrium messages/strategies.

The results of this chapter are also a contribution to the theory of mechanism design. In Sect. 5.2 we show that the multi-rate multicast problem with strategic users is the combination of a market and a public goods problem with strategic users. Such problems have not been previously investigated within the context of mechanism design.

5.1.4 Comparison with Related Work

Within the context of single rate and multi-rate multicast problems, studies have addressed issues of bandwidth/rate allocation [1–12], routing [8, 13–15] and reliability [16, 17]. Most of the literature on rate allocation is done via the notion of fairness [1, 2, 6, 7, 9], specifically max-min fairness [18] and proportional fairness [19]. The authors of [1] develop a unified framework for diverse fairness objectives via the notion of fair allocation of utilities. A more general approach to rate allocation is via utility maximization. Utility maximization is more general because rate allocation with the fairness property is utility maximizing when the utility has a special form. The authors of [10–12] investigated multi-rate multicast problems with a utility maximization objective.

In all the aforementioned papers, it is assumed the agents/users are not strategic, that is, they are price-takers who are willing to follow/obey the rules of the resource allocation mechanism.

In contrast to all the above papers, our work considers the multi-rate multicast problems with *strategic users*, that is, users which are self-utility maximizers, and do not necessarily obey the rules of the resource allocation mechanism, but have to be incentivized/induced to follow them. To the best of our knowledge, the mechanism, proposed in this chapter, is the first to present a mechanism possessing properties (P1)–(P3) for the multi-rate multicast service provisioning problem with strategic users.

5.1.5 Organization of the Chapter

The rest of the chapter is organized as follows. In Sect. 5.2 we formulate the multi-rate multicast service provisioning problem with strategic users. In Sect. 5.3 we describe the allocation mechanism/game form we propose for the solution of the multi-rate multicast service provisioning problem. In Sect. 5.4 we analyze the properties of the proposed mechanism. The proofs of all the results established in this chapter appear in Appendix B.

5.2 The Multi-Rate Multicast Problem with Strategic Network Users, Problem Formulation

In this Section we present the formulation of the multi-rate multicast problem in wired communication networks with strategic users. We proceed as follows, In Sect. 5.2.1 we formulate the centralized multi-rate multicast problem the solution of which we want to implement in Nash equilibria. In Sect. 5.2.2 we formulate the decentralized multi-rate multicast problem with strategic network users, and state our assumptions and objectives.

5.2.1 The Centralized Problem

We consider a wired network with N disjoint groups of *strategic* users; we denote the set of groups by $\mathcal{N} = \{G_1, G_2, \dots, G_N\}$. The network topology, the capacity of the network links, and the routes assigned to users' services are fixed and given. We denote user j in group G_i by (j, G_i). The utility function of user (j, G_i), $G_i \in \mathcal{N}$, has the form

$$V_{(j,G_i)}(x_{(j,G_i)}, t_{(j,G_i)}) = U_{(j,G_i)}(x_{(j,G_i)}) - t_{(j,G_i)}. \tag{5.1}$$

The term $U_{(j,G_i)}(x_{(j,G_i)})$ expresses user (j, G_i)'s *satisfaction* from the service $x_{(j,G_i)}$ it receives. The term $t_{(j,G_i)}$ represents the tax (money) user (j, G_i) pays for the services it receives. We assume that $U_{(j,G_i)}$ is a concave and increasing function of the service $x_{(j,G_i)}$ user (j, G_i) receives, and $t_{(j,G_i)} \in \mathbb{R}$. When $t_{(j,G_i)} > 0$ user (j, G_i) pays money for the services it receives; this money is paid to other network users. When $t_{(j,G_i)} < 0$ user (j, G_i) receives money from other users. Overall, the amount of money paid by some of the network users must be equal to the amount of money received by the rest of the users so that $\sum_{G_i \in \mathcal{N}} \sum_{j \in G_i} t_{(j,G_i)} = 0$.

Denote: by $\mathbf{L}$ the set of links of the network; by c_l the capacity of link l; by $\mathcal{R}_{(j,G_i)}$ the set of links l, $l \in \mathbf{L}$, that form the route of user (j, G_i), (as pointed out above each user's route is fixed); by $G_i(l)$ the set of users in G_i who use link l, i.e., $G_i(l) = \{j : j \in G_i \text{ and } l \in \mathcal{R}_{(j,G_i)}\}$; by $x_{G_i}(l)$ the maximum amount of bandwidth requested by group G_i at link l, i.e., $x_{G_i}(l) := \max_{j \in G_i(l)}\{x_{(j,G_i)}\}$; by ${G_i}^{\max}(l)$ the set of users in G_i using link l and request $x_{G_i}(l)$ amount of bandwidth, i.e., ${G_i}^{\max}(l) := \{(j, G_i) : x_{(j,G_i)} = x_{G_i}(l)\}$; by $(j, G_i^{\max}(l))$ a user in $G_i^{\max}(l)$; by $\mathbf{L}_{G_i}$ the set of links used by users in group G_i, i.e., $\mathbf{L}_{G_i} := \{l : \exists (j, G_i) \text{ s.t. } l \in \mathcal{R}_{(j,G_i)}\}$; by $\mathcal{R}^{\max}_{(j,G_i)}$ the set of links l, $l \in \mathcal{R}_{(j,G_i)}$, such that $x_{(j,G_i)} = x_{G_i}(l)$, i.e. $\mathcal{R}^{\max}_{(j,G_i)} = \{l : l \in \mathcal{R}_{(j,G_i)} \text{ s.t. } (j, G_i) = (j, G_i^{\max}(l))\}$; by Q_l the set of groups that include at least one user using link l, i.e., $Q_l := \{G_i : l \in \mathbf{L}_{G_i}\}$.

We assume that a central authority (the network manager) has access to all of the above information. The objective of this authority is to solve the following centralized optimization problem that we call **Max.0**

$$\max_{x,t} \sum_{G_i \in \mathcal{N}} \sum_{j \in G_i} \left[U_{(j,G_i)}(x_{(j,G_i)}) - t_{(j,G_i)} \right] \qquad \textbf{Max.0} \tag{5.2}$$

subject to

$$\sum_{G_i \in Q_l} \max_{j \in G_i(l)} x_{(j,G_i)} \le c_l, \qquad \forall\, l \in \mathbf{L}, \tag{5.3}$$

$$\sum_{G_i \in \mathcal{N}} \sum_{j \in G_i} t_{(j,G_i)} = 0, \tag{5.4}$$

$$x_{(j,G_i)} \ge 0, \qquad \forall j \in G_i, G_i \in \mathcal{N}, \tag{5.5}$$

where $(\boldsymbol{x}, \boldsymbol{t}) = (x_{(j,G_i)}, t_{(j,G_i)}, j \in G_i, G_i \in \mathcal{N})$. The inequalities in (5.3) express the capacity constraints that must be satisfied at each network link. The equality in (5.4) expresses the fact that the budget must be balanced, i.e., the total amount of money paid by some of the users must be equal to the amount of money received by the rest of the users. The inequalities in (5.5) express the fact that the users' received rates $x_{(j,G_i)}$, $G_i \in \mathcal{N}$, must be nonnegative. Every $(\boldsymbol{x}, \boldsymbol{t})$ that satisfies Eqs. (5.3)–(5.5) is called a *feasible* allocation/solution.

Problem **Max.0** is equivalent to problem **Max.1** below,

$$\max_{\boldsymbol{x}} \sum_{G_i \in \mathcal{N}} \sum_{j \in G_i} U_{(j,G_i)}(x_{(j,G_i)}) \qquad \textbf{Max.1} \tag{5.6}$$

subject to

$$\sum_{G_i \in Q_l} \sum_{(j,G_i) \in G_i(l)} x_{(j,G_i)} \leq c_l, \quad \forall j \in G_i(l), \forall\, l \in \mathbf{L}, \tag{5.7}$$

$$x_{(j,G_i)} \geq 0, \forall j \in G_i, \quad G_i \in \mathcal{N}, \tag{5.8}$$

in the following sense. The set of inequalities in (5.7) and (5.8) result in the same domain of solutions $\boldsymbol{x}$ as the set of inequalities in (5.3) and (5.5). Thus, any optimal solution $(x_{(j,G_i)}, j \in G_i, G_i \in \mathcal{N})$ of problem **Max.1** along with any $\boldsymbol{t} = \{t_{(j,G_i)}, j \in G_i, G_i \in \mathcal{N}\}$ such that $\sum_{G_i \in \mathcal{N}} \sum_{j \in G_i} t_{(j,G_i)} = 0$ is also an optimal solution $(x^*_{(j,G_i)}, t^*_{(j,G_i)}, j \in G_i, G_i \in \mathcal{N})$ of **Max.0**. We will refer to **Max.1** as the centralized multi-rate multicast problem.

Let $E(l)$ be the set of inequalities defined by (5.7) for link l. Every element of $E(l)$ is denoted by $e(l)$, $e(l) \in E(l)$. Define $E(l, (j, G_i)) \subseteq E(l)$ by

$$E(l, (j, G_i)) := \{e(l) \subseteq E(l) : x_{(j,G_i)} \text{ appears in } e(l)\}. \tag{5.9}$$

Let $\mathcal{U}$ denote the set of functions

$$U : \mathbb{R}_+ \cup \{0\} \rightarrow \mathbb{R}_+ \cup \{0\} \tag{5.10}$$

where U is concave and increasing, and $\mathbb{R}_+$ denotes the set of non-negative real numbers. Let $\mathbf{T}$ denote the set of all possible network topologies, network resources and user routes. Consider problem **Max.1** for all possible realizations

$$\times_{G_i \in \mathcal{N}} \times_{j \in G_i} U_{(j,G_i)} \times T \in \mathcal{U}^{\sum_{G_i \in \mathcal{N}} |G_i|} \times \mathbf{T}, \tag{5.11}$$

of the users' utilities, the network topology, its resources and the users' routes. Then the solution of **Max.1** for each (U, T) $\in \mathcal{U}^{\sum_{G_i \in \mathcal{N}} |G_i|} \times \mathbf{T}$ defines a map

$$\Gamma : \mathcal{U}^{\sum_{G_i \in \mathcal{N}} |G_i|} \times \mathbf{T} \rightarrow \mathcal{A}, \tag{5.12}$$

where $\mathcal{A} \in \mathbb{R}_+^{\sum_{G_i \in \mathcal{N}} |G_i|}$ is the set of all possible rate/bandwidth allocations to the network's users. We call Γ the solution of the centralized problem.

5.2.2 The Decentralized Problem with Strategic Users

We consider the network model of the previous section with the following assumptions on its information structure.

(A1) Each user knows *only* his own utility; this utility is his own private information. Each user also knows the *function space* $\mathcal{U}$ to which the utilities of all other users belong.
(A2) Each user behaves strategically, that is, each user is not a price-taker. The users's objective is to maximize his own utility function.
(A3) The network manager knows the topology and resources of the network. This knowledge is the manager's private information. The network manager is not a profit-maker (i.e. he does not have a utility function).
(A4) The network manager receives requests for service from the network users. Based on these requests, he announces to each user (j, G_i),

1. The multicast group to which the user belongs.
2. The set of links that form user (j, G_i)'s route, $\mathcal{R}_{(j,G_i)}$.
3. The capacity of each link in $\mathcal{R}_{(j,G_i)}$.

(A5) Based on the network manager's announcement, each strategic user competes for resources (bandwidth) at each link of his route with the other users in that link.[1]

From the above description it is clear that the information in the network is decentralized. Every user in each group only knows his own utility but does not know the other users' utilities or the network's topology and its resources. The network manager knows the network's topology and its resources, but does not know the users' utilities. It is also clear that the network manager (which is not profit maker) acts like an accountant who sets up the users' routes, specifies the users competing for resources/bandwidth at each link, collects the money from the users (j, G_i) that pay tax (i.e. $t_{(j,G_i)} > 0$) and distributes it to those users who receive money.

As a consequence of assumptions (A1)–(A5) we have at each link of the network a decentralized resource allocation problem which can be studied/analyzed within the context of implementation theory. These decentralized resource allocation problems are not *independent/decoupled*, as the rate that each user receives at any link of his own route must be the same. This constraint is dictated by the nature of the multi-rate

[1] Since in this chapter we present decentralized resource allocation mechanisms in equilibrium form, it is reasonable to assume that during the play of the game at each link $l \in \mathbf{L}$, each user of link l learns the set of the other users competing for bandwidth at l.

multicast service provisioning problem and has a direct implication on the nature of the mechanism/game form we present in Sect. 5.3.

Under the above assumptions the objective is to determine a game form/mechanism which has the following properties for each realization $(T, U_{(j,G_i)}, j \in G_i, G_i \in \mathcal{N})$:

(P1) It implements weakly in NE the social welfare maximizing correspondence defined by the centralized problem **Max.1**. (We note the social welfare maximizing correspondence is implementable in NE, cf. Sect. 2.1.4.)
(P2) It is individually rational, that is, the network users voluntarily participate in the decentralized bandwidth allocation process.
(P3) It is budget balanced at every NE of the game it induces, as well as at all off equilibrium messages/strategies that result in feasible allocations.

5.2.3 Key Features/Natures of the Problem

Multi-rate multicast service provisioning with strategic users is the **combination of a market problem and a public goods problem**. Thus, the model as well as the allocation problem are *new*, even within the context of the mechanism design. Specifically, resource allocation among groups is a market problem; resource allocation among the users of the same group is a public goods problem.

The market component: One can see that bandwidth allocation among groups is a market problem as follows. One can consider a group as a single agent. The demand of this group at each link of the network is the maximum of demands of the users of the group on that link. So, with each group considered a single agent/singleton the multi-rate multicast service provisioning problem with strategic users becomes equivalent to the unicast service provisioning problem with strategic users. It is shown, Chap. 3 of the thesis, that the unicast service provisioning problem with strategic users is a market problem. At each link, the price per unit of bandwidth paid collectively by each group[2] using the link is the same.

The public goods component: One can see that the resource allocation problem among the users of the same group is a public goods problem as follows. At equilibrium, the group receives at each link of the network a bandwidth/rate equal to the maximum requested by a user in the group. Each user of the group receives, in general, a different rate, and the members of the group that use the link must collectively pay the price per unit of bandwidth charged at the link. At each link, each user of a group using the link contributes, in general, a different percentage of the price per of unit of bandwidth charged at that link; this percentage depends on the amount of bandwidth received by the user, the user's utility, and the number of users that are present in the group and use the link. Consequently, the resource allocation problem along users of the same group is a public goods problem.

---

[2] The price per unit of bandwidth paid collectively by each multicast group at a link l is equal to the sum of the prices paid by the members' of the group who use the link l.

In the following two sections we present a mechanism/game form for the problem formulated in this section and prove that it possess properties (P1–P3) stated in Sect. 5.2.2.

5.3 A Mechanism for Rate Allocation

Based on the characteristics of the multi-rate multicast problem, we present guidelines for the design of rate allocation mechanisms in Sect. 5.3.1. In Sect. 5.3.2, we specify a mechanism/game form for the decentralized rate allocation problem formulated in Sect. 5.2. In Sect. 5.3.3, we discuss and interpret the components of the mechanism.

5.3.1 Guidelines for the Design of the Mechanism

In Sect. 5.2.3 we pointed out that the multi-rate multicast problem with strategic users is the combination of a market problem and a public goods problem. Therefore, the mechanism for rate allocation must capture both aspects/components of the problems. We now discuss the attributes a mechanism must have so that it can capture the market component and the public goods component of the multi-rate multicast problem.

To address the *market* characteristics of the problem the mechanism must be such that:

1. All groups that use a particular link must pay the same price per unit of bandwidth at the link.
2. The bandwidth allocation to groups at each link must satisfy the link's capacity constraint.
3. The budget must be balanced, that is the sum of payments of all the groups that use the network must be equal to zero at equilibrium and off equilibrium.

To address the *public goods* characteristics of the problem the mechanism must be such that:

4. At any link l, different users of the same group that use the link pay, in general, different prices per unit of bandwidth at link l. Specifically: if user a of group G requires more bandwidth than user b of group G at link l, user a must not pay less per unit of bandwidth at link l than user b. In general, if users a and b require the same amount of bandwidth at link l, they do not necessarily pay the same price per unit of bandwidth at l because they may have different utility functions.
5. The price that user i of group G pays per unit of bandwidth at a particular link that he uses must not be under his control; that is, the price must be determined by the messages/strategies of the other users that use the same link. This feature of the mechanism is a consequence of the users' strategic behavior.

With these considerations in mind we proceed to specify our mechanism.

5.3.2 Specification of the Mechanism

A game form/mechanism (c.f. Sect. 2.1.2) consists of two components $\mathcal{M}$, f. The component $\mathcal{M}$ denotes the users' *message/strategy space*, $\mathcal{M}$ defines the information the users are allowed to communicate with one another during the message exchange process. The component f is the *outcome function*; f defines for every message/strategy profile, the bandwidth/rate allocated to each user and the tax (subsidy) each user pays (receives).

For the decentralized resource allocation problem formulated in Sect. 5.2 we propose a game form/mechanism the components of which we describe below.

Message space: The message/strategy space for user (j, G_i), $j \in G_i$, $G_i \in \mathcal{N}$, is given by $\mathcal{M}_{(j,G_i)} = \mathbb{R}_+^{|\mathcal{R}_{(j,G_i)}|+1}$. Specifically, a message of user j is of the form

$$\boldsymbol{m}_{(j,G_i)} = \left[x_{(j,G_i)}, \pi_{(j,G_i)}^{l_{j_1}}, \pi_{(j,G_i)}^{l_{j_2}}, \ldots, \pi_{(j,G_i)}^{l_{j|\mathcal{R}_{(j,G_i)}|}} \right],$$

where $|\mathcal{R}_{(j,G_i)}|$ denotes the number of links along the route $\mathcal{R}_{(j,G_i)}$. The component $x_{(j,G_i)}$ denotes the bandwidth/rate user (j, G_i) requests at all the links of his route. The component $\pi_{(j,G_i)}^{l_{jk}} \in [0, \Upsilon]$,[3] $0 \leq \Upsilon < \infty$, $k = 1, 2, \ldots, |\mathcal{R}_{(j,G_i)}|$, denotes the price per unit of bandwidth user (j, G_i) is willing to pay at link l_{jk} of his route.

Remark 5.1 Due to the nature of the multi-rate multicast service provisioning problem (see Sect. 5.2) the bandwidth/rate allocated to any user (j, G_i), $j \in G_i$, $G_i \in \mathcal{N}$, must be the same at all links of his route. Thus, the nature of message $\boldsymbol{m}_{(j,G_i)}$ is a consequence of the above requirement.

Outcome Function: The outcome function f

$$f : \times_{G_i \in \mathcal{N}} \times_{j \in G_i} \mathcal{M}_{(j,G_i)} \to \mathbb{R}_+^{\sum_{G_i \in \mathcal{N}} |G_i|} \times \mathbb{R}^{\sum_{G_i \in \mathcal{N}} \sum_{j \in G_i} |\mathcal{R}_{(j,G_i)}|} \tag{5.13}$$

is defined as follows: for any

$$\boldsymbol{m} := (\boldsymbol{m}_{i \in G_1}, \boldsymbol{m}_{j \in G_2}, \ldots, \boldsymbol{m}_{k \in G_N}) \in \mathcal{M} := \times_{G_i \in \mathcal{N}} \times_{j \in G_i} \mathcal{M}_{(j,G_i)},$$

$$\begin{aligned} f(\boldsymbol{m}) &= f(\boldsymbol{m}_{i \in G_1}, \boldsymbol{m}_{j \in G_2}, \ldots, \boldsymbol{m}_{k \in G_N}) \\ &= \left((x_{(i,G_1)}, \boldsymbol{t}_{(i,G_1)})_{i \in G_1}, (x_{(j,G_2)}, \boldsymbol{t}_{(j,G_2)})_{j \in G_2}, \ldots, (x_{(k,G_N)}, \boldsymbol{t}_{(k,G_N)})_{k \in G_N} \right), \end{aligned} \tag{5.14}$$

where $\boldsymbol{t}_{(j,G_i)} := (t_j^{l_{j1}}, t_j^{l_{j2}}, \ldots, t_j^{l_{j|\mathcal{R}_{j,G_i}|}})$, for every (j, G_i), $j \in G_i$, $G_i \in \mathcal{N}$, is the tax (subsidy) that user (j, G_i) pays (receives) to (from) the other users, through

[3] For technical reasons (cf. Theorem 5) we choose Υ to be arbitrary and large but finite.

the network manager, for each link $l_{j_k} \in \mathcal{R}_{(j,G_i)}$, and $x_{(j,G_i)}$, $j \in G_i, G_i \in \mathcal{N}$, represents the amount of bandwidth/rate allocated to user (j, G_i).

The tax $t_j^{l_{j_k}}, k = 1, 2, \ldots, |\mathcal{R}_{(j,G_i)}|, \forall j \in G_i, G_i \in \mathcal{N}$, is defined in accordance with the number of multicast groups using link l. We consider four cases.

- **Case A**. $|Q_l| = 1$

Let $Q_l = \{G_\varsigma\}$. Then, for any $j \in G_\varsigma(l)$,

$$t^l_{(j,G_\varsigma)} = \mathbb{I}\{x_{(j,G_\varsigma)} = x_{G_\varsigma}(l)\}\left\{0 \cdot \mathbb{I}\{x_{G_\varsigma}(l) \le c_l\} + \frac{1\{x_{G_\varsigma}(l) > c_l\}}{1 - 1\{x_{G_\varsigma}(l) > c_l\}}\right\}. \quad (5.15)$$

The function $\mathbb{I}\{\cdot\}$ denotes the indicator function, i.e.,

$$\mathbb{I}\{A\} = \begin{cases} 1 \text{ if } A \text{ holds;} \\ 0 \text{ otherwise.} \end{cases}$$

The function $1\{A\}$, used throughout the chapter, is defined as follows

$$1\{A\} = \begin{cases} 1 - \epsilon \text{ if } A \text{ holds;} \\ 0 \qquad \text{otherwise.} \end{cases}$$

where ϵ is bigger than zero and sufficiently small;[4] ϵ is chosen by the mechanism designer.

- **Case B**. $|Q_l| = 2$

Let $Q_l = \{G_\varsigma, G_{\varsigma+1}\}$. We consider two subcases, $|G_\varsigma{}^{\max}(l)| \ge 2$ and $|G_\varsigma{}^{\max}(l)| = 1$.

Part BI: $|G_\varsigma{}^{\max}(l)| \ge 2$.

Let the label of (j, G_ς) in $G^{\max}_\varsigma(l)$ be $(k, G^{\max}_\varsigma(l))$. Then:
If $(j, G_\varsigma) \in G^{\max}_\varsigma(l)$,

$$\begin{aligned} t^l_{(k,G_\varsigma)} &= \pi_{(k+1,G_\varsigma{}^{\max}(l))} x_{(j,G_\varsigma)} + \frac{\left(P_{G_\varsigma{}^{\max}(l)} - P_{G_{\varsigma+1}{}^{\max}(l)}\right)^2}{\alpha |G_\varsigma{}^{\max}(l)|} \\ &\quad - 2\frac{P_{G_{\varsigma+1}(l)^{\max}}}{|G_\varsigma{}^{\max}(l)|}\left[P_{G_\varsigma{}^{\max}(l)} - P_{G_{\varsigma+1}{}^{\max}(l)}\right]\left[\frac{x_{G_{\varsigma+1}}(l) + x_{(j,G_\varsigma)} - c_l}{\gamma}\right] \\ &\quad + \frac{1\{x_{(j,G_\varsigma)} > 0\}1\{x_{G_{\varsigma+1}}(l) + x_{(j,G_\varsigma)} - c_l > 0\}}{1 - 1\{x_{(j,G_\varsigma)} > 0\}1\{x_{G_{\varsigma+1}}(l) + x_{(j,G_\varsigma)} - c_l > 0\}} \end{aligned} \quad (5.16)$$

If $(k, G_\varsigma) \notin G^{\max}_\varsigma(l)$ then

$$t^l_{(k,G_\varsigma)} = 0, \quad (5.17)$$

[4] Therefore, when A and B (both) hold, then $\frac{1\{A\}1\{B\}}{1-1\{A\}1\{B\}} \approx \frac{1}{0^+}$ is well defined and it becomes a large number.

where α and γ are sufficiently large constants, $P_{G_\zeta^{\max}(l)} = \sum_{j \in G_\zeta{}^{\max}(l)} \pi_{(j,G_\zeta{}^{\max}(l))}$, and $k+1$ is defined mod $(|G_\zeta^{\max}|)$.

Part BII: If $|G_\zeta{}^{\max}(l)| = 1$. Then:
If $(j, G_\zeta) \in G_\zeta^{\max}(l)$,

$$t^l_{(j,G_\zeta)} = P_{G_{\zeta+1}^{\max}(l)} x_{(j,G_\zeta)} + \frac{\left(P_{G_\zeta{}^{\max}(l)} - P_{G_{\zeta+1}{}^{\max}(l)}\right)^2}{\alpha} - 2P_{G_{\zeta+1}{}^{\max}}(l)\left[P_{G_\zeta{}^{\max}(l)} - P_{G_{\zeta+1}{}^{\max}(l)}\right]\left[\frac{x_{G_{\zeta+1}}(l) + x_{(j,G_\zeta)} - c_l}{\gamma}\right] + \frac{1\{x_{(j,G_\zeta)} > 0\}1\{x_{G_{\zeta+1}}(l) + x_{(j,G_\zeta)} - c_l > 0\}}{1 - 1\{x_{(j,G_\zeta)} > 0\}1\{x_{G_{\zeta+1}}(l) + x_{(j,G_\zeta)} - c_l > 0\}} \tag{5.18}$$

If $(j, G_\zeta) \notin G_\zeta^{\max}(l)$ then

$$t^l_{(j,G_\zeta)} = 0. \tag{5.19}$$

- **Case C**. $|Q_l| = 3$

Let $Q_l = \{G_\zeta, G_{\zeta+1}, G_{\zeta+2}\}$. We consider two subcases,$|G_\zeta{}^{\max}(l)| \geq 2$ and $|G_\zeta{}^{\max}(l)| = 1$.

Part CI: $|G_\zeta{}^{\max}(l)| \geq 2$. Then:
Let the label of (j, G_ζ) in $G_\zeta^{\max}(l)$ be $(k, G_\zeta^{\max}(l))$. Then:
If $(j, G_\zeta) \in G_\zeta^{\max}(l)$,

$$t^l_{(j,G_\zeta)} = \pi_{(k+1,G_\zeta{}^{\max}(l))} x_{(j,G_\zeta)} + \frac{\left(P_{G_\zeta{}^{\max}(l)} - P_{-G_\zeta{}^{\max}(l)}\right)^2}{\alpha\,|G_\zeta{}^{\max}(l)|} - 2\frac{P_{-G_\zeta{}^{\max}}(l)}{|G_\zeta{}^{\max}(l)|}\left[P_{G_\zeta{}^{\max}(l)} - P_{-G_\zeta{}^{\max}(l)}\right]\left[\frac{\mathcal{E}_{-G_\zeta{}^{\max}(l)} + x_{(j,G_\zeta)}}{\gamma}\right] + \frac{1\{x_{(j,G_\zeta)} > 0\}1\{\mathcal{E}_{-G_\zeta^{\max}(l)} + x_{(j,G_\zeta)} > 0\}}{1 - 1\{x_{(j,G_\zeta)} > 0\}1\{\mathcal{E}_{-G_\zeta^{\max}(l)} + x_{(j,G_\zeta)} > 0\}} \tag{5.20}$$

If $(j, G_\zeta) \notin G_\zeta^{\max}(l)$ then

$$t^l_{(j,G_\zeta)} = 0. \tag{5.21}$$

Part CII: $|G_\zeta{}^{\max}(l)| = 1$.
If $(j, G_\zeta) \in G_\zeta^{\max}(l)$,

$$t^l_{(j,G_\varsigma)} = P_{-G_\varsigma{}^{\max}(l)} x_{(j,G_\varsigma)} - 2P_{-G_\varsigma{}^{\max}} \left[\pi_{(j,G_\varsigma{}^{\max}(l))} - P_{-G_\varsigma{}^{\max}} \right] \times \left[\frac{\mathcal{E}_{-G_\varsigma{}^{\max}(l)} + x_{(j,G_\varsigma)}}{\gamma} \right] + \frac{\left(\pi_{(j,G_\varsigma{}^{\max}(l))} - P_{-G_\varsigma{}^{\max}(l)}\right)^2}{\alpha} + \frac{1\{x_{(j,G_\varsigma)} > 0\} 1\{\mathcal{E}_{-G^{\max}_\varsigma(l)} + x_{(j,G_\varsigma)} > 0\}}{1 - 1\{x_{(j,G_\varsigma)} > 0\} 1\{\mathcal{E}^l_{-G^{\max}_\varsigma(l)} + x_{(j,G_\varsigma)} > 0\}} \tag{5.22}$$

If $(j, G_\varsigma) \notin G^{\max}_\varsigma(l)$ then

$$t^l_{(j,G_\varsigma)} = 0, \tag{5.23}$$

where

$$\mathcal{E}_{-G_\varsigma{}^{\max}(l)} := x_{G_{\varsigma+1}}(l) + x_{G_{\varsigma+2}}(l) - c_l,$$
$$P_{G_\varsigma{}^{\max}(l)} := \sum_{j \in G_\varsigma{}^{\max}(l)} \pi_{(j,G_\varsigma{}^{\max}(l))},$$
$$P_{-G_\varsigma{}^{\max}(l)} := \frac{P_{G_{\varsigma+1}{}^{\max}(l)} + P_{G_{\varsigma+2}{}^{\max}(l)}}{2}.$$

- **Case D**. $|Q_l| > 3$

  Let $G_i \in Q_l$. We consider two subcases, $|G_i{}^{\max}(l)| \geq 2$ and $|G_i{}^{\max}(l)| = 1$.
  Part DI: $|G_i{}^{\max}(l)| \geq 2$.

Let the label of (j, G_ς) in $G^{\max}_\varsigma(l)$ be $(k, G^{\max}_\varsigma(l))$. Then:
If $(j, G_\varsigma) \in G^{\max}_\varsigma(l)$,

$$t^l_{(j,G_i)} = \pi_{(k+1,G_i{}^{\max}(l))} x_{(j,G_\varsigma)} + \frac{\left(P_{G_i{}^{\max}(l)} - P_{-G_i{}^{\max}(l)}\right)^2}{|G_i{}^{\max}(l)|} - 2\frac{P_{-G_i{}^{\max}}}{|G_i{}^{\max}(l)|} \left[P_{G_i{}^{\max}(l)} - P_{-G_i{}^{\max}(l)} \right] \left[\frac{\mathcal{E}_{-G_i{}^{\max}(l)} + x_{(j,G_i)}}{\gamma} \right] + \frac{1\{x_{(j,G_i)} > 0\} 1\{\mathcal{E}_{-G^{\max}_i(l)} + x_{(j,G_i)} > 0\}}{1 - 1\{x_{(j,G_i)} > 0\} 1\{\mathcal{E}_{-G^{\max}_i(l)} + x_{(j,G_i)} > 0\}} + \frac{\Gamma^l_{G_i}}{|G^{\max}_i(l)|} \tag{5.24}$$

If $(j, G_\varsigma) \notin G^{\max}_\varsigma(l)$ then

$$t^l_{(j,G_\varsigma)} = 0. \tag{5.25}$$

where $\mathcal{E}_{-G_i{}^{\max}(l)}$, $P_{G_i{}^{\max}(l)}$, and $P_{-G_i{}^{\max}(l)}$ are defined by equations similar to (5.28)–(5.30).

Part DII: $|G_i{}^{\max}(l)| = 1$.

If $(j, G_\zeta) \in G_\zeta^{\max}(l)$,

$$
\begin{aligned}
t^l_{(j,G_i)} &= P_{-G_i{}^{\max}} x_{j,G_i} + \left(\pi_{(j,G_i{}^{\max}(l))} - P_{-G_i{}^{\max}(l)}\right)^2 \\
&\quad - 2P_{-G_i{}^{\max}}\left[\pi_{(j,G_i{}^{\max}(l))} - P_{-G_i{}^{\max}}\right]\left[\frac{\mathcal{E}_{-G_i{}^{\max}(l)} + x_{(j,G_i)}}{\gamma}\right] \\
&\quad + \frac{1\{x_{(j,G_i)} > 0\}1\{\mathcal{E}_{-G_i^{\max}(l)} + x_{(j,G_i)} > 0\}}{1 - 1\{x_{(j,G_i)} > 0\}1\{\mathcal{E}_{-G_i^{\max}(l)} + x_{(j,G_i)} > 0\}} + \Gamma^l_{G_i}
\end{aligned}
\tag{5.26}
$$

If $(j, G_\zeta) \notin G_\zeta^{\max}(l)$ then

$$
t^l_{(j,G_\zeta)} = 0, \tag{5.27}
$$

where,

$$
\mathcal{E}_{-G_i{}^{\max}(l)} := \Big\{ \sum_{\substack{G_k \in Q_l \\ G_k \neq G_i}} x_{G_k}(l) \Big\} - c_l, \tag{5.28}
$$

$$
P_{G_i{}^{\max}(l)} := \sum_{j \in G_i{}^{\max}(l)} \pi_{(j,G_i{}^{\max}(l))}, \tag{5.29}
$$

$$
P_{-G_i{}^{\max}(l)} := \frac{\sum_{\substack{G_k \in Q_l \\ G_k \neq G_i}} P_{G_k{}^{\max}(l)}}{|Q_l| - 1} = \frac{\sum_{\substack{G_k \in Q_l \\ G_k \neq G_i}} \sum_{j \in G_k{}^{\max}(l)} \pi_{(j,G_k{}^{\max}(l))}}{|Q_l| - 1}, \tag{5.30}
$$

$$
\begin{aligned}
\Gamma^l_{G_i} &:= \frac{\sum_{\substack{G_s \in Q_l \\ G_s \neq G_i}} \sum_{\substack{G_r \in Q_l \\ G_r \neq G_i, G_s}} \left(2P_{G_s{}^{\max}(l)} P_{G_r{}^{\max}(l)}\left(1 + \frac{x_{G_s}(l)}{\gamma}\right)\right)}{(|Q_l| - 1)(|Q_l| - 2)} \\
&\quad + \frac{2\sum_{\substack{G_s \in Q_l \\ G_s \neq G_i}} \sum_{\substack{G_r \in Q_l \\ G_r \neq G_i, G_s}} \sum_{\substack{G_t \in Q_l \\ G_t \neq G_i, G_s, G_r}} P_{G_r{}^{\max}(l)}\left(P_{G_s{}^{\max}(l)}\mathcal{E}_{G_t{}^{\max}(l)} - P_{G_t{}^{\max}(l)} x_{G_s}(l)\right)}{(|Q_l| - 1)^2(|Q_l| - 3)\gamma} \\
&\quad + \frac{2\sum_{\substack{G_s \in Q_l \\ G_s \neq G_i}} \sum_{\substack{G_r \in Q_l \\ G_r \neq G_i, G_s}} P_{G_r{}^{\max}(l)}\left(P_{G_s{}^{\max}(l)}\mathcal{E}_{G_r{}^{\max}(l)} - P_{G_r{}^{\max}(l)} x_{G_s}(l)\right)}{(|Q_l| - 1)^2(|Q_l| - 2)\gamma} \\
&\quad - \frac{2P^2_{-G_i{}^{\max}(l)}\,\mathcal{E}_{-G_i{}^{\max}(l)}}{\gamma} - \frac{\sum_{\substack{G_s \in Q_l \\ G \neq G_i}} {P_{G_s{}^{\max}(l)}}^2}{(|Q_l| - 1)} - P^2_{-G_i{}^{\max}(l)}.
\end{aligned}
\tag{5.31}
$$

Next we specify additional subsidies $\mathcal{S}^l$ that user (j, G_i), $j \in G_i$, $G_i \in \mathcal{N}$, may receive. For that matter we consider all links $l \in \mathbf{L}$ where $|Q_l| \leq 3$. For each such link l, we define the quantity

$$\mathcal{S}^l := \sum_{G_\varsigma \in Q_l} \sum_{(j,G_\varsigma) \in G_\varsigma{}^{\max}(l)} -t^l_{(j,G_\varsigma)} \Big[\mathbb{I}\{\text{Case B}\} + \mathbb{I}\{\text{Case C}\} \Big]. \tag{5.32}$$

Since α and γ are sufficiently large,

$$\begin{aligned} \mathcal{S}^l &= o(1) - \sum_{G_\zeta, G_\varsigma \in Q_l} P_{G_\varsigma^{\max}(l)} x_{G_\varsigma}(l) \Big[\mathbb{I}\{\text{Case B(Part BI)}\} + \mathbb{I}\{\text{Case C(Part CI)}\} \Big] \\ &\quad - \sum_{G_\zeta, G_\varsigma \in Q_l} P_{-G_\varsigma^{\max}(l)} x_{G_\varsigma}(l) \Big[\mathbb{I}\{\text{Case B(Part BII)}\} + \mathbb{I}\{\text{Case C(Part CII)}\} \Big] \\ &:= o(1) - S^l_+. \end{aligned} \tag{5.33}$$

For each $l \in \mathbf{L}$ where $|Q_l| \leq 3$ the network manager chooses at random a user $k_l \notin \bigcup_{G_i \in Q_l} G_i$ and assigns the subsidy $\mathcal{S}^l$ to user k_l. Let $l_1, l_2, \ldots, l_r$ be the set of links such that $|Q_{l_i}| \leq 3$, and let k_{l_i} be the corresponding users that receive $\mathcal{S}^{l_i}$.

Based on the above, the tax (subsidy) paid (received) by user (j, G_i), $j \in G_i$, $G_i \in \mathcal{N}$, is the following. If $(j, G_i) \neq k_{l_1}, k_{l_2}, \ldots k_{l_r}$ then

$$t_{(j,G_i)} = \sum_{l \in \mathcal{R}_{(j,G_i)}} t^l_{(j,G_i)}, \tag{5.34}$$

where for each $l \in \mathcal{R}_{(j,G_i)}$, $t^l_{(j,G_i)}$ is determined in accordance with $|Q_l|$. If $(j, G_i) = k_{l_n}$ for some $k_{l_n} \in \bigcup_{m=1}^{r} k_{l_m}$, then

$$t_{k_{l_i}} = \sum_{l \in \mathcal{R}_{k_{l_i}}} t^l_{k_{l_i}} + \mathcal{S}^{l_i}, \tag{5.35}$$

where $\mathcal{S}^{l_i}$ is defined by (5.32) and $\mathcal{R}_{k_{l_i}}$ is the set of links used by user k_{l_i}.

Note that $\mathcal{S}^{l_i}$ is not controlled by user k_{l_i}. Thus, the presence (or absence) of $\mathcal{S}^{l_i}$ does not influence the strategic behavior of user k_{l_i}. We have assumed here that the users $k_{l_1}, k_{l_2}, \cdots, k_{l_r}$, are distinct. Expressions similar to the above hold when the users $k_{l_1}, k_{l_2}, \cdots, k_{l_r}$ are not distinct.

5.3.3 Discussion/Interpretation of the Mechanism

We now interpret the mechanism presented in Sect. 5.3.2, based on the guidelines for its design, presented in Sect. 5.3.1. We focus on Case D (Part DI). The mechanism's interpretation is similar in all other cases. To proceed with the interpretation we define:

$$\Delta_1^{(j,G_i)}(l) := \pi_{(j+1,G_i{}^{\max}(l))} x_{(j,G_i)},$$

$$\Delta_2^{(j,G_i)}(l) := \frac{\left(P_{G_i{}^{\max}(l)} - P_{-G_i{}^{\max}(l)}\right)^2}{|G_i{}^{\max}(l)|} - 2\frac{P_{-G_i{}^{\max}}(l)}{|G_i{}^{\max}(l)|}\left[P_{G_i{}^{\max}(l)} - P_{-G_i{}^{\max}(l)}\right]\left[\frac{\mathcal{E}_{-G_i{}^{\max}(l)} + x_{(j,G_i)}}{\gamma}\right] + \frac{1\{x_{(j,G_i)} > 0\}1\{\mathcal{E}_{-G_i^{\max}(l)} + x_{(j,G_i)} > 0\}}{1 - 1\{x_{(j,G_i)} > 0\}1\{\mathcal{E}_{-G_i^{\max}(l)} + x_{(j,G_i)} > 0\}}$$

$$\Delta_3^{(j,G_i)}(l) := \frac{\Gamma_{G_i}^l}{|G_i^{\max}(l)|}$$

$$\Delta_4^{(j,G_i)}(l) := \mathbb{I}\{x_{(j,G_i)} = x_{G_i}(l)\}.$$

Note that (5.24) and (5.25) can be collectively rewritten as follows,

$$t_{(j,G_i)}^l = \left[\Delta_1^{(j,G_i)}(l) + \Delta_2^{(j,G_i)}(l) + \Delta_3^{(j,G_i)}(l)\right] \times \Delta_4^{(j,G_i)}(l). \tag{5.36}$$

$\Delta_1^{(j,G_i)}(l)$, $\Delta_2^{(j,G_i)}(l)$, $\Delta_3^{(j,G_i)}(l)$, and $\Delta_4^{(j,G_i)}(l)$ collectively represent the tax (subsidy) user (j, G_i) pays (receives) for using link l. The terms $\Delta_1^{(j,G_i)}(l)$ and $\Delta_4^{(j,G_i)}(l)$ (respectively, $\Delta_2^{(j,G_i)}(l)$ and $\Delta_3^{(j,G_i)}(l)$) capture/describe the public goods (respectively, market) component of the problem.

We begin with the interpretation of the public goods terms. Note that user (j, G_i) pays taxes (receives subsidies) at link l only if his bandwidth demand is the maximum among the users of group G_i at link l. This is expressed by the term $\Delta_4^{(j,G_i)}(l)$. By assumption that the cardinality of the set of users from G_i who have maximum bandwidth demand at link l is greater than one. Assume now that (j, G_i) is one of the users of group G_i that have maximum bandwidth demand at link l, and let $(k, G_i^{\max}(l))$ be the index of this user in $G_i^{\max}(l)$. The price per unit of bandwidth at link l that this user pays is not under his control; it is determined by the message/strategy ($\pi_{(k+1,G_i{}^{\max}(l))}$) of user $(k+1, G_i^{\max}(l))$, that is user $k+1$ of the group $G_i^{\max}(l)$.[5] This is reflected in the term $\Delta_1^{(j,G_i)}(l)$ which represents the amount of tax user (j, G_i) pays for the bandwidth he receives at link l. The two terms are consistent with the design guidelines associated with the public goods features of the mechanism presented in Sect. 5.3.1. Specifically, terms $\Delta_1^{(j,G_i)}(l)$ and $\Delta_4^{(j,G_i)}(l)$ demonstrate that: (i) at any link l, if user a of group G_i receives more bandwidth than user b of the same group, then user a pays no less for this bandwidth than user b; (ii) if two users a and b of the same group require maximum amount of bandwidth at link l they do not necessarily pay the same price per unit of bandwidth at that link.

[5] The situation where (j, G_i) is the only user of group G_i with the maximum demand at link l is discussed in other cases (e.g. Case D (Part DII)), where it is shown again that the price user (j, G_i) pays per unit of bandwidth at link l is not controlled by him.

As a result of the specification and interpretation of the terms $\Delta_1^{(j,G_i)}(l)$ and $\Delta_4^{(j,G_i)}(l)$, the price group G_i pays per unit of bandwidth at link l is the sum of the prices its users with maximum demand at link l pay. That is,

$$P_{G_i^{\max}(l)} = \sum_{(j,G_i^{\max}(l)) \in G_i^{\max}(l)} \pi_{(j,G_i^{\max}(l))}.$$

We continue with interpretation of the market terms of the tax function. The term $\Delta_2^{(j,G_i)}(l)$ provides the following incentives to the groups using link l: (1) To bid/propose the same price per unit of bandwidth at that link. (2) To collectively request a total bandwidth that does not exceed the capacity of the link. The incentive provided to all groups to bid the same price per unit of bandwidth is described by the term $\frac{\left(P_{G_i{}^{\max}} - P_{-G_i{}^{\max}}\right)^2}{|G_i{}^{\max}(l)|}$. The incentive provided to all users to collectively request a total bandwidth that does not exceed the link's capacity is captured by the term

$$\frac{1\{x_{(j,G_i)} > 0\}1\{\mathcal{E}_{-G_i^{\max}(l)} + x_{(j,G_i)} > 0\}}{1 - 1\{x_{(j,G_i)} > 0\}1\{\mathcal{E}^l_{-G_i^{\max}(l)} + x_{(j,G_i)} > 0\}}.$$

Note that each group is very heavily penalized if it requests a nonzero bandwidth at l, and, collectively, all the groups using l request a total bandwidth that exceeds the link's capacity c_l. A joint incentive provided to all users to bid the same price per unit of bandwidth and to utilize the total capacity of the link is captured by the term

$$-2\frac{P_{-G_i{}^{\max}}(l)}{|G_i{}^{\max}(l)|}\left[P_{G_i{}^{\max}(l)} - P_{-G_i{}^{\max}(l)}\right]\left[\frac{\mathcal{E}_{-G_i{}^{\max}(l)} + x_{(j,G_i)}}{\gamma}\right].$$

The goal of the term $\Delta_3^{(j,G_i)}(l)$ is to lead to a balanced budget. It is important to note that the term $\Delta_3^{(j,G_i)}(l)$ is not controlled by group G_i, consequently, by any user in group G_i. Therefore, the presence of $\Delta_3^{(j,G_i)}(l)$ does not affect the behavior of any user of group G_i. The terms $\Delta_2^{(j,G_i)}(l)$ and $\Delta_3^{(j,G_i)}(l)$ are consistent with the guidelines that were presented in Sect. 5.3.1 concerning the market features of the mechanism.

5.4 Properties of the Mechanism

We prove that the mechanism proposed in Sect. 5.3 has the following properties. (P1) It implements the solution of problem **Max.0** in Nash equilibria. (P2) It is individually rational. (P3) It is budget-balanced at every feasible allocation.

We establish the above properties by proceeding as follows. First, we prove that the game induced by the mechanism proposed in Sect. 5.3 has at least one pure NE

(Theorem 5.2), and that all NE of the game induced by the game form/mechanism of Sect. 5.3 result in feasible solutions of the centralized problem **Max.0** (Theorem 5.3). Afterwards, we establish that the mechanism is budget-balanced at all feasible allocations. (Lemma 5.6). Then, we show that network users voluntarily participate in the allocation process. We do this by showing that each user's utility/payoff resulting from the allocations corresponding to all NE of the game induced by the mechanism is greater than or equal to zero, the payoff each user receives when he does not participate in the allocation process (Theorem 5.7). Finally, we show that the mechanism implements in Nash equilibria the solution of the centralized allocation problem **Max.0** (Theorem 5.8).

We present the proofs of the following theorems and lemmas in Appendix B.

Theorem 5.2 (EXISTENCE OF NE) *The game induced by the mechanism of Sect. 5.3 has at least one pure NE.*

Theorem 5.3 (FEASIBILITY) *If $\boldsymbol{m}^* = (\boldsymbol{x}^*, \boldsymbol{\pi}^*)$ is a NE of the game induced by the game form of Sect. 5.3, then the allocation $\boldsymbol{x}^*$ is a feasible solution of problem* ***Max.0****.*

The following lemma presents some key properties of NE prices and rates.

Lemma 5.4 *Let $\boldsymbol{m}^* = (\boldsymbol{x}^*, \boldsymbol{\pi}^*)$ be a NE of the game induced by game form of Sect. 5.3. Then for every $l \in \boldsymbol{L}$ and $G_i \in Q_l$, we have,*

$$P^*_{-G_i{}^{\max}(l)} = P^*_{G_i{}^{\max}(l)} =: P^*_{G^{\max}(l)} \quad \forall\, G_i \in Q_l \tag{5.37}$$

$$P^*_{G^{\max}(l)} \left[\frac{\mathcal{E}^*_{-G_i{}^{\max}(l)} + x^*_{G_i}(l)}{\gamma} \right] = 0. \tag{5.38}$$

For every user $(j, G_i{}^{\max}(l))$ where $G_i \in Q_l$, we have,

$$\left. \frac{\partial t^l_{(j,G_i^{\max}(l))}}{\partial x_{G_i}(l)} \right|_{\boldsymbol{m}=\boldsymbol{m}^*} = \begin{cases} \pi^*_{(j+1,G_i{}^{\max}(l))}, & \text{if } |G_i{}^{\max}(l)| \geq 2 \text{ and } G_i \in Q_l, \\ P^*_{G^{\max}(l)}, & \text{otherwise,} \end{cases} \tag{5.39}$$

and

$$\left[\frac{\partial U_{(j,G_i^{\max}(l))}(x_{(j,G_i^{\max}(l))})}{\partial x_{(j,G_i^{\max}(l))}} - \sum_{l \in \mathcal{R}^{\max}_{(j,G_i)}} \frac{\partial t^l_{(j,G_i^{\max}(l))}}{\partial x_{(j,G_i^{\max}(l))}} \right]_{\boldsymbol{m}=\boldsymbol{m}^*} = 0. \tag{5.40}$$

An immediate consequence of Lemma 5.4 and the specification of the tax for each user, defined by Eqs. (5.16)–(5.35), is the following.

Corollary 5.5 *At every NE $\boldsymbol{m}^*$ of the mechanism the tax function has the following form,*

$$t^l_{(j,G_i^{\max}(l))}(\boldsymbol{m}^*) = \begin{cases} \pi^*_{(j+1,G_i^{\max}(l))} x^*_{G_i}(l) & \textit{Case B, Part BI;} \\ P^*_{G^{\max}} x^*_{G_i}(l) & \textit{Case B, Part BII;} \\ \pi^*_{(j+1,G_i^{\max}(l))} x^*_{G_i}(l) & \textit{Case C, Part CI;} \\ P^*_{G^{\max}} x^*_{G_i}(l) & \textit{Case C, Part CII;} \\ \pi^*_{(j+1,G_i^{\max}(l))} x^*_{G_i}(l) - \frac{P^*_{G^{\max}(l)} x^*_{-G_i}(l)}{|G_i^{\max}|} & \textit{Case D, Part DI;} \\ P^*_{G^{\max}(l)} \left(x^*_{G_i}(l) - x^*_{-G_i}(l) \right) & \textit{Case D, Part DII} \end{cases} \tag{5.41}$$

where

$$x^*_{-G_i}(l) = \frac{\sum_{\substack{G_j \\ G_j \neq G_i \\ G_j \in Q_l}} x^*_{G_j}(l)}{|Q_l| - 1}.$$

When $(j, G_i) \notin G_i^{\max}(l)$, $t^l_{(j,G_i)}(\boldsymbol{m}^*) = 0$. *Therefore,*

$$t_{(j,G_i)}(\boldsymbol{m}^*) = \sum_{l \in \mathcal{R}^{\max}_{(j,G_i)}} t^l_{(j,G_i)}(\boldsymbol{m}^*), \tag{5.42}$$

for $(j, G_i) \neq k_{l_1}, k_{l_2}, \ldots, k_{l_r}$, *(cf. Sect. 5.3), and for* $j = k_{l_s}$, $s = 1, 2, \ldots, r$,

$$t_{(j,G_i)}(\boldsymbol{m}^*) = \sum_{l \in \mathcal{R}^{\max}_{(j,G_i)}} t^l_{(j,G_i)}(\boldsymbol{m}^*) - S^{*j}_{+} \tag{5.43}$$

In the following lemma, we prove that the proposed mechanism is always budget balanced.

Lemma 5.6 *The proposed mechanism/game form is budget balanced at every feasible allocation. That is, the mechanism is budget-balanced at all allocations corresponding to NE messages as well as to all off-equilibrium messages/strategies that result in feasible allocations.*

The next result asserts that the mechanism/game form proposed in Sect. 5.3 is individually rational.

Theorem 5.7 (INDIVIDUAL RATIONALITY): *The game form specified in Sect. 5.3 is individually rational, i.e., at every NE the corresponding allocation* $(\boldsymbol{x}^*, \boldsymbol{t}^*)$ *is weakly preferred by all users to zero, the payoff each user receives when he does not participate in the allocation process.*

In the following theorem we show that every NE of the game induced by the game form proposed in Sect. 5.3 is efficient.

Theorem 5.8 (Nash Implementation): *The allocation* $(f(\boldsymbol{m}^*) = (\boldsymbol{x}^*, \boldsymbol{t}^*))$ *corresponding to a NE message* $\boldsymbol{m}^*$ *is an optimal solution of the centralized problem* ***Max.0****.*

References

1. Sarkar S, Tassiulas L (2002) Fair allocation of utilities in multirate, multicast networks: a framework for unifying diverse fairness objectives. IEEE Trans Autom Control 47(6):933–944
2. Sarkar S, Tassiulas L (2002) A framework for routing and congestion control for multicast information flows. IEEE Trans Inform Theory 48(10):2690–2708
3. Sarkar S, Tassiulas L (2006) Layered multicast rate control based on lagrangian relaxation and dynamic programming. IEEE J Sel Areas Commun 24(8):1464–1474
4. Kar K, Sarkar S, Tassiulas L (2002) A scalable low-overhead rate control for multi-rate multicast sessions. IEEE J Sel Areas Commun 20(8):1541–1557
5. Kar K, Tassiulas L (2002) Layered multicast rate control based on Lagrangian relaxation and dynamic programming. IEEE J Sel Areas Commun 24(8):1464–1474
6. Rubenstein D, Kurose J, Towsley D (1999) The impact of multicast layering on network fairness. In: Proceedings of ACM-SIGCOMM, University of Massachusetts, Cambridge, 1999
7. Graves E, Srikant R, Towsley D (1999) Decentralized computation of weighted max-min fair bandwidth allocation in networks with multicast flows. In: Proceedings of International Workshop on Digital Communications (IWDC)
8. Shapiro J, Towsley D, Kurose J (2000) Optimization-based congestion control for multicast communications. In: Procedings of INFOCOM
9. Tzeng H, Siu K (1997) On max-min fair congestion for multicast abr service in atm. IEEE J Sel areas Commun 15(3):545–556
10. Deb D, Srikant R (2004) Congestion control for fair resource allocation in networks with multicast flows. IEEE/ACM Trans Netw 12(2):261–273
11. Stoenescu T, Liu M, Teneketzis D (2007) Multi-rate multicast service provisioning, part i: an algorithm for optimal price splitting along multicast trees. Math Methods Oper Res 65(2):199–228
12. Stoenescu T, Liu M, Teneketzis D (2007) Multirate multicast service provisioning, part ii: a tatonnement process for rate allocation. Math Methods Oper Res 65(3):389–415
13. Zegura E (1993) Routing algorithms in multicast switching topologies. In: Proceedings of Allerton conference on communication, control and computing
14. Park W, Owen H, Zegura E (1993) Sonet/sdh multicast routing algorithms in symmetrical three-stage networks. In: Proceedings of international, communication conference (ICC)
15. Donahoo M, Zegura E (1997) Center selection and migration for wide-area multicast routing. J High Speed Netw 6(2):141–164
16. Gupta R, Walrand J (1999) Average bandwidth and delay for reliable multicast. In: Proceedings of IFIP performance
17. Duffield N, Horowitz J, Towsley D, Wei W, Friedman T (2002) Multicast-based loss inference with missing data. IEEE J Sel areas Commun 20(4):700–713
18. Bertsekas DP, Gallager RG (1992) Data networks. Prentice-Hall, Englewood
19. Kelly F (1997) Charging and rate control for elastic traffic. Euro Trans Telecommun 8(1):33–37

Chapter 6
Summary and Future Directions

6.1 Summary

In this thesis we investigated static decentralized resource allocation problems with strategic users. Unicast service provisioning and multi-rate multicast service provisioning problems arise in wired communication networks; power allocation and spectrum sharing arise in wireless communication networks. The models associated with unicast service provisioning, power allocation and spectrum sharing, and multi-rate multicast service provisioning capture generic issues that arise in market problems, public goods problems, and problems that are a combination of markets and public goods, respectively.

For each of the problems arising in wired networks we developed game forms/mechanisms and analyzed them in equilibrium. We proved that the proposed game forms possess the following properties. (P1) They implement in NE the social welfare maximizing correspondence. (P2) They are budget balanced at the allocations corresponding to all NE of the game induced by the mechanism, as well as at all feasible allocations corresponding to off equilibrium messages. (P3) They are individually rational, that is users voluntarily participate in the allocation process specified by the mechanism. For the problem arising in wireless networks we developed a game form that possesses properties (P2) and (P3), and implements in NE the Pareto correspondence.

Within the context of the above mentioned network problems, the game forms developed in this thesis are the only currently existing mechanisms that possess all the above-stated properties. The results on power allocation and spectrum sharing, as well as the results on multi-rate multicast service provisioning are also a contribution to the state of the art of implementation theory.

There are several problems of paramount importance that remain unsolved and are worthy of investigation. Below we discuss some of these problems.

A. Kakhbod, *Resource Allocation in Decentralized Systems with Strategic Agents*, Springer Theses, DOI: 10.1007/978-1-4614-6319-1_6,
© Springer Science+Business Media New York 2013

6.2 Future Directions

6.2.1 Algorithmic Issues

Currently, we do not have algorithms (tâtonnement processes/iterative processes) for the computation of the NE of the games induced by the game forms we developed. The lack of algorithms for decentralized resource allocation problems where strategic users posses private information is a major open problems in implementation theory. The major difficulty in constructive algorithms that guarantee to converge to NE is the following. Consider an algorithm for a decentralized resource allocation problem where strategic users possess private information. At each stage of the algorithm each user updates his strategy/message. Since the users' utilities are not common knowledge, after each update a strategic user, say user i, can report any strategy he deems to be advantageous for himself; that is, user i can misreport/misrepresent his update and the other users can not check whether or not user i is following the rules of the algorithm. Consequently, the algorithm must provide incentives to the users/agents to follow its rules at each one of its stages. Such a provision of incentives must be based on all the information available at the current stage, and must, in general, take the whole future into account. We have not been able to discover an algorithm with the above feature. To the best of our knowledge, algorithms with the above feature are not currently available.

6.2.2 Dynamic Environments

In this thesis we focused on static decentralized resource allocation problems where the system characteristics (e.g. the network topology, the number of users, the users' utilities) do not change with time. The development of mechanisms (that is, situations where the network topology and resources, and/or the number of users, and/or the users' utilities vary with time) is an important open problem. The dynamic mechanisms currently available in the literature [1–3] are direct game forms/direct revelation mechanisms, and the existing results are on truthful implementation, which does not guarantee that for any environment all NE of the game induced by the direct revelation mechanism result in allocations that are in the choice set of the social choice rule/goal correspondence (see [4]). In our opinion, progress in the design of decentralized resource allocation mechanisms for dynamic environments will require a better understanding of the interplay between implementation theory and dynamic game theory. We also believe that resolving the key issues associated with the development of algorithms for static decentralized resource allocation problems (cf. Sect. 6.2.1) will help us understand better the nature of dynamic decentralized resource allocation problems where strategic users possess private information.

6.2.3 Beyond Quasi-Linear Forms

In this thesis, within the context of unicast and multi-rate multicast service provisioning we addressed resource allocation problems where the users' utilities are quasi-linear. In many real systems the network objective or the users' utilities are not separable in money (tax). Problems with non-quasi-linear objectives are harder to solve as they do not have a general structure or methodology for their solution, and have not received much attention in the mechanism design literature. Developing game forms/mechanisms that implement in some equilibrium concept non-quasi linear network objectives is a problem of fundamental importance. A step in this direction are the results reported in the power allocation and spectrum sharing problem we investigated in Chap. 4.

References

1. Athey S, Segal I (2007) An efficient dynamic mechanism, preprint
2. Bergemann D, Valimaki J (2010) The dynamic pivot mechanism. Econometrica 78(2):771–789
3. Pavan A, Segal I, Toikka J (2011) Dynamic mechanism design: Revenue equivalence, profit maximization and information disclosure, preprint
4. Dasgupta P, Hammond P, Maskin E (1979) The implementation of social choice rules: some general results on incentive compatibility. Rev Econ Stud 46(2):185–216

Appendix A
Appendix for Unicast Service Provisioning

Proof of Theorem 3.1. By the construction of the mechanism $x_i^* \geq 0$ for all $i \in \mathcal{N}$. Suppose that $\mathrm{x}^* = (x_1^*, \ldots, x_N^*)$ is such that the capacity constraint is violated at some link l and $x_j^* > 0$ (i.e. user j will be heavily charged because $\frac{1\{x_j^*>0\}1\{\mathcal{E}_{-j}^{*l}+x_j^*>0\}}{1-1\{x_j^*>0\}1\{\mathcal{E}_{-j}^{*l}+x_j^*>0\}} \approx \frac{1}{0^+}$ which is a large number). Now, Consider x_j such that: (i) either $x_j > 0$ and $\sum_{\substack{k \in \mathcal{G}^l \\ k \neq j}} x_k^* + x_j \leq c_l$; or (ii) $x_j = 0$. Then, $\frac{1\{x_j>0\}1\{\mathcal{E}_{-j}^{*l}+x_j>0\}}{1-1\{x_j>0\}1\{\mathcal{E}_{-j}^{*l}+x_j>0\}} = 0$, therefore,

$$\mathcal{V}_j(m_j, m_{-j}^*) > \mathcal{V}_j(m_j^*, m_{-j}^*), \tag{A.1}$$

and (A.1) contradicts the fact that $m^* = (m_j^*, m_{-j}^*)$ is a NE. Consequently, x^* is a feasible allocation of problem **Max**. □

Proof of Lemma 3.2. We prove this lemma by considering the case $|\mathcal{G}^l| > 3$. The cases $|\mathcal{G}^l| = 2$ and $|\mathcal{G}^l| = 3$ can be proved similarly.
Consider user $i \in \mathcal{G}^l$ ($|\mathcal{G}^l| > 3$). Since user i does not control Φ_i^l defined by (3.14), (i.e. Φ_i^l does not depend on x_i and p_i^l),

$$\frac{\partial \Phi_i^l}{\partial x_i} = \frac{\partial \Phi_i^l}{\partial p_i^l} = 0. \tag{A.2}$$

Equation (3.13) along with (A.2) imply

$$\frac{\partial t_i^l}{\partial p_i^l}|_{m=m^*} = 2\Big[(p_i^{*l} - P_{-i}^{*l}) - P_{-i}^{*l}\Big(\frac{\mathcal{E}_{-i}^{*l} + x_i^*}{\gamma}\Big)\Big] = 0. \tag{A.3}$$

A. Kakhbod, *Resource Allocation in Decentralized Systems with Strategic Agents*, Springer Theses, DOI: 10.1007/978-1-4614-6319-1,
© Springer Science+Business Media New York 2013

Summing Eq. (A.3) over all $i \in \mathcal{G}^l$, we get,

$$\sum_{i \in \mathcal{G}^l} \frac{\partial t_i^l}{\partial p_i^l}|_{m=m^*} = \sum_{i \in \mathcal{G}^l} \left[(p_i^{*l} - P_{-i}^{*l}) - P_{-i}^{*l} \Big(\frac{\mathcal{E}_{-i}^{*l} + x_i^*}{\gamma} \Big) \right]$$

$$= \sum_{i \in \mathcal{G}^l} - P_{-i}^{*l} \Big(\frac{\mathcal{E}_{-i}^{*l} + x_i^*}{\gamma} \Big) = 0, \tag{A.4}$$

which, because of Theorem 3.1 and the positivity of prices, implies

$$- P_{-i}^{*l} \Big(\frac{\mathcal{E}_{-i}^{*l} + x_i^*}{\gamma} \Big) = 0. \tag{A.5}$$

for every $i \in \mathcal{G}^l$. Then Eq. (A.5) gives

$$p_i^{*l} = P_{-i}^{*l}. \tag{A.6}$$

for all $i \in \mathcal{G}^l$. From Eqs. (A.5) and (A.6) it follows that,

$$p^{*l} \Big(\frac{\mathcal{E}^{*l}}{\gamma} \Big) = 0, \tag{A.7}$$

$$p_i^{*l} = p_j^{*l} = P_{-i}^{*l} = p^{*l}. \tag{A.8}$$

Equations (A.7) and (A.8) along with (3.13) give

$$\frac{\partial t_i^l}{\partial x_i}|_{m=m^*} = p^{*l}. \tag{A.9}$$

By (A.7)–(A.9)[1] the proof is complete. □

Proof of Lemma 3.3. Equation (3.30) together with (3.31) and (3.32) imply that $\sum_{l \in \mathbf{L}} \sum_{i \in \mathcal{G}^l} t_i^{*l} = 0$. Now, we prove that the proposed mechanism is also budget balanced off equilibrium. First we show that, for every $l \in \mathbf{L}$ where $|\mathcal{G}^l| > 3$

$$\sum_{i \in \mathcal{G}^l, |\mathcal{G}^l| > 3} t_i^l = 0. \tag{A.10}$$

[1] Note that, since the derivative of an indicator function is a Dirac delta function ([1], p. 94), to have a well defined derivative of t_i with respect to x_i at the boundary, i.e., when $\sum_{i \in \mathcal{G}^l} x_i = c_l$, the differentiation is from the left. This observation holds throughout the proofs appearing in this Appendix.

By a little algebra we can show the following equalities,

$$\sum_{i\in\mathcal{G}^l} p_i^{l2} = \sum_{i\in\mathcal{G}^l}\left[\frac{\sum_{\substack{j\in\mathcal{G}^l\\ j\neq i}} p_j^{l\,2}}{|\mathcal{G}^l|-1}\right],$$

$$\sum_{i\in\mathcal{G}^l}\left(2p_i^l P_{-i}^l + 2P_{-i}^l p_i^l\frac{x_i}{\gamma} - P_{-i}^l x_i\right) = \sum_{i\in\mathcal{G}^l}\left[\frac{\sum_{\substack{j\in\mathcal{G}^l\\ j\neq i}}\sum_{\substack{k\in\mathcal{G}^l\\ k\neq i,j}}\left(2p_j^l p_k^l(1+\frac{x_j}{\gamma}) - x_j p_k^l\right)}{(|\mathcal{G}^l|-1)(|\mathcal{G}^l|-2)}\right],$$

$$\sum_{i\in\mathcal{G}^l} P_{-i}^l p_i^l\frac{\mathcal{E}_{-i}^l}{\gamma} = \sum_{i\in\mathcal{G}^l}\left[\frac{\sum_{\substack{j\in\mathcal{G}^l\\ j\neq i}}\sum_{\substack{k\in\mathcal{G}^l\\ k\neq i,j}}\sum_{\substack{r\in\mathcal{G}^l\\ r\neq i,j,k}} 2p_k^l p_j^l\mathcal{E}_r^l}{\gamma(|\mathcal{G}^l|-1)^2(|\mathcal{G}^l|-3)} + \frac{\sum_{\substack{j\in\mathcal{G}^l\\ j\neq i}}\sum_{\substack{k\in\mathcal{G}^l\\ k\neq i,j}} 2p_k^l p_j^l\mathcal{E}_k^l}{\gamma(|\mathcal{G}^l|-1)^2(|\mathcal{G}^l|-2)}\right],$$

$$\sum_{i\in\mathcal{G}^l} P_{-i}^{l\;2}\frac{x_i}{\gamma} = \sum_{i\in\mathcal{G}^l}\left[\frac{\sum_{\substack{j\in\mathcal{G}^l\\ j\neq i}}\sum_{\substack{k\in\mathcal{G}^l\\ k\neq i,j}}\sum_{\substack{r\in\mathcal{G}^l\\ r\neq i,j,k}} x_j p_r^l}{\gamma(|\mathcal{G}^l|-1)^2(|\mathcal{G}^l|-3)} + \frac{\sum_{\substack{j\in\mathcal{G}^l\\ j\neq i}}\sum_{\substack{k\in\mathcal{G}^l\\ k\neq i,j}} x_j p_k^l}{\gamma(|\mathcal{G}^l|-1)^2(|\mathcal{G}^l|-2)}\right].$$

From the above equalities we conclude that

$$\sum_{i\in\mathcal{G}^l}\left[P_{-i}^l x_i + (p_i^l - P_{-i}^l)^2 - 2P_{-i}^l(p_i^l - P_{-i}^l)\left(\frac{\mathcal{E}_{-i}^l + x_i}{\gamma}\right)\right] = -\sum_{i\in\mathcal{G}^l}\Phi_i^l \quad \text{(A.11)}$$

Equation (A.11) along with Eq. (3.13) imply that $\sum_{i\in\mathcal{G}^l,|\mathcal{G}^l|>3} t_i^l = 0$.

Next consider all links $l \in \mathbf{L}$ where $|\mathcal{G}^l| = 2, 3$. In accordance with the notation in Sect. 3.3, let these links be $l_1, l_2, \ldots, l_r$. Then, by the specification of the tax function (cf. Sect. 3.3.1, Eqs. (3.6)–(3.10)) we obtain,

$$\sum_{j=1}^{r}\left\{\left[t_{i_{l_{j,1}}}^{l_j} + t_{i_{l_{j,2}}}^{l_j}\right]1\{|\mathcal{G}^{l_j}|=2\} + \left[t_{i_{l_{j,1}}}^{l_j} + t_{i_{l_{j,2}}}^{l_j} + t_{i_{l_{j,3}}}^{l_j}\right]1\{|\mathcal{G}^{l_j}|=3\}\right\} + \sum_{j=1}^{r} Q^{l_j} = 0, \quad \text{(A.12)}$$

where if $|\mathcal{G}^{l_j}| = 2$ then $\{i_{l_{j,1}}, i_{l_{j,2}}\} = \mathcal{G}^{l_j}$ and if $|\mathcal{G}^{l_j}| = 3$ then $\{i_{l_{j,1}}, i_{l_{j,2}}, i_{l_{j,3}}\} = \mathcal{G}^{l_j}$, $j = 1, 2, \ldots, r$.
Finally note that,

$$\sum_{i=1}^{N} t_i = \sum_{l\in\mathbf{L}:|\mathcal{G}^l|=2}\sum_{i\in\mathcal{G}^l} t_i^l + \sum_{l\in\mathbf{L}:|\mathcal{G}^l|=3}\sum_{i\in\mathcal{G}^l} t_i^l + \sum_{l\in\mathbf{L}:|\mathcal{G}^l|>3}\sum_{i\in\mathcal{G}^l} t_i^l + \sum_{j=1}^{r} Q^{l_j} = 0. \quad \text{(A.13)}$$

□

Proof of Theorem 3.4. We need to show that $\mathcal{V}_i(\mathrm{x}^*, t_i^*) \geq \mathcal{V}_i(0, 0) = 0$ for every $i \in \mathcal{N}$. By the property of NE it follows that

$$\mathcal{V}_i(\mathrm{x}^*, t^*) \geq \mathcal{V}_i(\mathrm{x}_{-i}^*, x_i, t_i, t_{-i}^*) \qquad \forall (x_i, \mathbf{t}_i). \tag{A.14}$$

So, it is enough to find $(x_i, \mathrm{p}_i) \in \mathcal{M}_i$ such that

$$\mathcal{V}_i(\mathrm{x}_{-i}^*, x_i, t_i, t_{-i}^*) \geq 0. \tag{A.15}$$

We set $x_i = 0$ and examine the cases $|\mathcal{G}^l| = 2$, $|\mathcal{G}^l| = 3$ and $|\mathcal{G}^l| > 3$, separately.

- CASE 1, $|\mathcal{G}^l| = 2$
  With $x_i = 0$, $p_j^l = p^{*l}$ and $x_j = x_j^*$, Eq. (3.6) defines the following function $\mathcal{F}_2(p_i^l)$:

$$\mathcal{F}_2(p_i^l) := \frac{(p_i^l - p^{*l})^2}{\alpha} - 2^{*l}(p_i^l - p^{*l})\left(\frac{x_j^* - c_l}{\gamma}\right)$$

  Clearly, at

$$p_i^l = p^{*l} \tag{A.16}$$

  $\mathcal{F}_2(p^{*l}) = 0$. Then, from Eq. (3.6) it follows that

$$t_i^l(\mathrm{x}_{-i}^*, 0, \mathrm{p}_{-i}^*, p^{*l}) = 0. \tag{A.17}$$

- CASE 2, $|\mathcal{G}^l| = 3$
  Denote by i, j, k the users of link l. With $x_i = 0$, $x_j = x_j^*$, $x_k = x_k^*$ and $p_j^l = p_k^l = p^{*l}$, Eq. (3.8) defines the following function $\mathcal{F}_3(p_i^l)$:

$$\begin{aligned}\mathcal{F}_3(p_i^l) &:= (p_i^l - p^{*l})^2 - 2p^{*l}(p_i^l - p^{*l})\left(\frac{x_j^* + x_k^* - c_l}{\gamma}\right) + \Omega_i^{*l} \\ &= p_i^{l2} - 2p_i^l p^{*l}\left(1 + \frac{\mathcal{E}_{-i}^{*l}}{\gamma}\right) + p^{*l2}\left(1 + 2\frac{\mathcal{E}_{-i}^{*l}}{\gamma}\right) + \Omega_i^{*l}\end{aligned} \tag{A.18}$$

  $\mathcal{F}_3(p_i^l)$ is a quadratic polynomial in p_i^l. Setting $\mathcal{F}_3(p_i^l) = 0$ we obtain the root

$$\wp_{i,3}^l = p^{*l}\left(1 + \frac{\mathcal{E}_{-i}^{*l}}{\gamma}\right) + \sqrt{\left(p^{*l}\frac{\mathcal{E}_{-i}^{*l}}{\gamma}\right)^2 + x_{-i}^* p^{*l} + \frac{p^{*l2}(c_l - \mathcal{E}_{-i}^{*l})}{\gamma}} \tag{A.19}$$

  Since by its definition γ is sufficiently large, it follows from Eq. (A.19) that $\wp_{i,3}^l > 0$, i.e. $\wp_{i,3}^l$ is a feasible price. Therefore, from Eq. (3.8) we obtain

$$t_i^l(\mathrm{x}_{-i}^*, 0, \mathrm{p}_{-i}^*, \wp_{i,3}^l) = 0. \tag{A.20}$$

- CASE 3, $\left|\mathcal{G}^l\right| > 3$
  With $x_i = 0, x_j = x_j^* \; \forall j \neq i, j \in \mathcal{G}^l, p_j^l = p^{*l}, j \in \mathcal{G}^l$, Eqs. (3.13) and (3.14) define (after a little algebra) the following function $\mathcal{F}_{>3}(p_i^l)$,

$$\mathcal{F}_{>3}(p_i^l) := (p_i^l - p^{*l})^2 - 2p^{*l}(p_i^l - p^{*l})\Big(\frac{\mathcal{E}_{-i}^{*l}}{\gamma}\Big) + \Phi_i^{*l}$$
$$= p_i^{l2} - 2p_i^l p^{*l}\Big(1 + \frac{\mathcal{E}_{-i}^{*l}}{\gamma}\Big) + p^{*l2}\Big(1 + 2\frac{\mathcal{E}_{-i}^{*l}}{\gamma}\Big) - x_{-i}^* p^{*l} \tag{A.21}$$

$\mathcal{F}_{>3}(p_i^l)$ is a quadratic polynomial in p_i^l. Setting $\mathcal{F}_{>3}(p_i^l) = 0$ we obtain the root

$$\wp_{i,>3}^l = p^{*l}\Big(1 + \frac{\mathcal{E}_{-i}^{*l}}{\gamma}\Big) + \sqrt{\Big(p^{*l}\frac{\mathcal{E}_{-i}^{*l}}{\gamma}\Big)^2 + x_{-i}^* p^{*l}} \tag{A.22}$$

where

$$x_{-i}^* := \frac{\sum_{j\neq i} x_j^*}{|\mathcal{G}^l| - 1}. \tag{A.23}$$

Since by its definition γ is sufficiently large, it follows from Eq. (A.22) that $\wp_{i,>3}^l > 0$ (i.e. $\wp_{i,>3}^l$ is a feasible price). Therefore, from Eq. (3.13) we get

$$t_i^l(\mathrm{x}_{-i}^*, 0, \mathrm{p}_{-i}^*, \wp_{i,>3}^l) = 0. \tag{A.24}$$

Consequently, at $m_i = (x_i, \mathrm{p}_i) = (0, p_i^{l_{i_1}}, p_i^{l_{i_2}}, \ldots, p_i^{l_{i_{|\mathcal{R}_i|}}})$, (where, $p_i^{l_{i_k}}, k = 1, 2, \ldots, |\mathcal{R}_i|$, are defined either by (A.16) or (A.19) or (A.22), depending on the cardinality $\mathcal{G}^{l_{i_k}}, k = 1, 2, \ldots, |\mathcal{R}_i|$), we obtain

$$\mathcal{V}_i(x, \mathbf{t})\Big|_{m=(m_i, m_{-i}^*)} = u_i(0) - \sum_{k=0}^{|\mathcal{R}_i|} t_i^{l_{i_k}}(\mathrm{x}_{-i}^*, 0, \mathrm{p}_{-i}^{*l_{i_k}}, p_i^{l_{i_k}})$$
$$= u_i(0) = 0. \tag{A.25}$$

when $i \neq k_{l_1}, k_{l_2}, \ldots, k_{l_r}$.

When $i = k_{l_j}, j = 1, 2, \ldots, r$,

$$\mathcal{V}_i(x, \mathbf{t})|_{m=(m_i, m_{-i}^*)} = u_i(0) - \sum_{k=0}^{|\mathcal{R}_i|} t_i^{l_{i_k}}(\mathrm{x}_{-i}^*, 0, \mathrm{p}_{-i}^{*l_{i_k}}, p_i^{l_{i_k}}) - \mathcal{Q}^{*l_j}$$
$$= -\mathcal{Q}^{*l_j} \geq 0, \tag{A.26}$$

where $\mathcal{Q}^{*l_j} = \mathcal{Q}^{*\{l_j:|\mathcal{G}^{l_j}|=2\}}$ or $\mathcal{Q}^{*l_j} = \mathcal{Q}^{*\{l_j:|\mathcal{G}^{l_j}|=3\}}$. Combining (A.14), (A.25) and (A.26) we obtain

$$\mathcal{V}_i(x_i^*, \mathbf{t}^*) \geq \mathcal{V}_i(x, \mathbf{t})\Big|_{m=(m_i, m_{-i}^*)} \geq 0 \tag{A.27}$$

and this establishes (A.15) and completes the proof. □

Proof of Theorem 3.5. Let $(\mathrm{x}^*, \mathrm{p}^*)$ be an arbitrary NE of the game $(\mathcal{M}, f, \mathcal{V})$ induced by the proposed game form. Then by the properties of NE, we must have that for every user $i \in \mathcal{N}$,

$$\frac{\partial \mathcal{V}_i(m)}{\partial x_i}|_{m=m^*} = \left[\frac{\partial u_i(x_i)}{\partial x_i} - \frac{\partial t_i(m)}{\partial x_i}\right]|_{m=m^*} = 0. \tag{A.28}$$

By Lemma 3.2, Eq. (A.28) is equivalent to

$$\frac{\partial u_i(x_i)}{\partial x_i} - \sum_{l \in \mathcal{R}_i} p^{*l} = 0. \tag{A.29}$$

Furthermore, by Lemma 3.2 we have $p^{*l}\mathcal{E}^{*l}/\gamma = 0$ and since $\gamma > 0$

$$p^{*l}\mathcal{E}^{*l} = p^{*l}\Big[\sum_{k \in \mathcal{G}^l} x_k^* - c_l\Big] = 0. \tag{A.30}$$

Equation (A.28) holds for every user $i \in \mathcal{N}$; Eq. (A.30) holds for every link $l \in \mathbf{L}$.

Consider now the centralized problem **Max**. Since the functions $u_i, i \in \mathcal{N}$ are concave and differentiable and the constraints are linear, Slater's condition [2] is satisfied, the duality gap is equal to zero, and the Karush Kuhn Tucker (KKT) conditions are necessary and sufficient to guarantee the optimality of any allocation $\mathrm{x} := (x_1, x_2, \ldots, x_N)$ that satisfies them. Let λ^l be the Lagrange multiplier corresponding to the capacity constraint for link l and ν_i be the Lagrange multiplier corresponding to the demand constraint. The Lagrangian for problem **Max** is

$$\mathcal{L}(x, \lambda, \nu) = \sum_{i \in \mathcal{N}} u_i(x_i) - \sum_{l \in \mathbf{L}} \lambda^l \Big(\sum_{i \in \mathcal{G}_l} x_i - c_l\Big) + \sum_{i \in \mathcal{N}} \nu_i x_i \tag{A.31}$$

and the KKT conditions are:

$$\frac{\partial \mathcal{L}(x^*, \lambda^*, \nu^*)}{\partial x_i} = \frac{\partial u_i(x_i^*)}{\partial x_i} - \sum_{l \in \mathcal{R}_i} \lambda^{*l} + \nu_i^* = 0 \tag{A.32}$$

$$\lambda^{*l}\Big(\sum_{i \in \mathcal{G}^l} x_i^* - c_l\Big) = 0 \quad \forall\, l \in \mathbf{L} \tag{A.33}$$

$$\nu_i^* x_i^* = 0 \quad \forall i \in \mathcal{N} \tag{A.34}$$

Since the KKT conditions are necessary and sufficient to guarantee the optimality of any allocation $\mathbf{x} = (x_1, x_2, \ldots, x_N)$ that satisfies them, it is enough to find ν_i^* and $\lambda^{*l}, l \in \mathbf{L}$, such that Eqs. (A.32)–(A.34) are satisfied.

Set $\nu_i^* = 0, i \in \mathcal{N}$, and $\lambda^{*l} = p^{*l}, l \in \mathbf{L}$. Then (A.34) is satisfied and (A.32) and (A.33) become,

$$\frac{\partial u_i(x_i^*)}{\partial x_i} - \sum_{l \in \mathcal{R}_i} p^{*l} = 0 \tag{A.35}$$

$$p^{*l}\Big(\sum_{i \in \mathcal{G}^l} x_i - c_l\Big) = 0 \quad \forall\, l \in \mathbf{L} \tag{A.36}$$

respectively, and they are satisfied because they are identical to Eqs. (A.28) and (A.30), respectively. Furthermore, by the construction of the game form $\sum_{i=1}^{N} t_i^* = 0$. Consequently, the solution $\mathrm{x}^* = (x_1^*, x_2^*, \ldots, x_N^*)$ of (A.35) and (A.36) along with the specification of $t_i^*, i = 1, 2, \ldots, N$, are an optimal solution of Problem **Max**. At the same time (A.35) and (A.36) and $\sum_{i=1}^{N} t_i^* = 0$ are satisfied by the allocation $f(m^*)$ corresponding to the NE m^*. Consequently, the NE m^* results in allocation $f(m^*) = (x_1^*, x_2^*, \ldots, x_N^*, t_1^*, t_2^*, \ldots, t_N^*)$ that is an optimal solution of Problem **Max**. Since the NE m^* was arbitrarily chosen, every NE m^* of the game form proposed in Sect. 3.3 results in an optimal solution of Problem **Max**. □

Proof of Theorem 3.6. First we note that an optimal solution

$$(\mathbf{x}^*, \mathbf{t}^*) = (x_1^*, x_2^*, \ldots, x_N^*, t_i^*, t_2^*, \ldots, t_N^*)$$

(where $t_i, i = 1, 2, \ldots, N$, are defined in Sect. 3.3.1) of Problem **Max** exists. This follows from: (i) the fact that each $u_i, i \in \mathcal{N}$, is concave and the space of the constraints described by Eqs. (3.3) and (3.4) is convex; (ii) the fact that $\sum_{i=1}^{N} t_i^* = 0$ by the construction of the game form. The KKT conditions for problem **Max** result in the following equations,

$$\frac{\partial u_i(x_i^*)}{\partial x_i^*} - \sum_{l \in \mathcal{R}_i} \lambda^{*l} + \nu_i^* = 0 \quad \text{(N equations)} \tag{A.37}$$

$$\lambda^{*l}(\sum_{i \in \mathcal{G}^l} x_i^* - c_l) = 0 \quad \text{(L equations)} \tag{A.38}$$

$$\nu_i^* x_i^* = 0 \quad \text{(N equations)} \tag{A.39}$$

We have $N + L + N$ equations in $L + N$ unknowns, $\lambda^{*l}, l \in \mathbf{L}$ and $\nu_i^*, i \in \mathcal{N}$. In general we have multiple solutions.

We want to show that for every solution $(\lambda^{*l}, \nu_i^*, l = 1, 2, \ldots, L, i = 1, 2, \ldots, N)$ of Eqs. (A.37)–(A.39) the message $\bar{m} = (\bar{m}_1, \bar{m}_2, \ldots, \bar{m}_N), \bar{m}_i = (\bar{x}_i, \bar{p}_i^l :$

$l \in \mathcal{R}_i$) with $\bar{x}_i = x_i^*$ and $\bar{p}_i^l = \lambda^{*l}$ for all $i \in \mathcal{N}$ and $l \in \mathcal{R}_i$, is a Nash equilibrium of the game induced by the proposed game form.

For that matter we note that by the selection of $\bar{m}$ we have

$$p_i^{*l} = p_j^{*l} = \lambda^{*l} = p^{*l} \tag{A.40}$$

for every i and $j \in \mathcal{G}^l$. By (A.38) and (A.40)

$$p^{*l}\Big(\sum_{j \in \mathcal{G}^l} x_j^* - c_l\Big) = \lambda^{*l}\Big(\sum_{j \in \mathcal{G}^l} x_j^* - c_l\Big) = 0 \tag{A.41}$$

and by (3.30) we obtain

$$\frac{\partial t_i^{*l}}{\partial x_i^*} = p^{*l} = \lambda^{*l} \tag{A.42}$$

for every $l \in \mathcal{R}_i$ and every $i \in \mathcal{N}$. Therefore, the message $\bar{m}$ satisfies all the conditions of Lemma 3.2.

Next we show that for every $i \in \mathcal{N}$, $\bar{m}_i$ is a solution of the problem,

$$\max_{m_i \in \mathcal{M}_i} \Big\{u_i(x_i) - \sum_{l \in \mathcal{R}_i} t_i^l(\bar{m}_{-i}, m_i)\Big\}$$

$$\text{subject to}$$

$$x_i \geq 0,\ p_i^l \geq 0 \qquad \forall l \in \mathcal{R}_i. \tag{A.43}$$

Any maximizing solution of (A.43) must satisfy

$$\frac{\partial u_i(x_i)}{\partial x_i} - \sum_{l \in \mathcal{R}_i} \frac{\partial t_i^l(\bar{m}_{-i}, m_i)}{\partial x_i} + r_i = 0 \tag{A.44}$$

$$\frac{\partial u_i(x_i)}{\partial p_i^l} - \sum_{l \in \mathcal{R}_i} \frac{\partial t_i^l(\bar{m}_{-i}, m_i)}{\partial p_i^l} + q_i^l = 0 \tag{A.45}$$

$\forall\, l \in \mathcal{R}_i$, where r_i and q_i^l are the Lagrange multipliers associated with the constraints $x_i \geq 0$, and $p_i^l \geq 0$, $l \in \mathcal{R}_i$, respectively. We set $r_i = \nu_i^*$ and $q_i^l = 0$ for every $l \in \mathcal{R}_i$. At $m_i = \bar{m}_i$, Eq. (A.44) is satisfied because of Eq. (A.37).

Furthermore at $m_i = \bar{m}_i$ Eq. (A.45) is satisfied since

$$\frac{\partial V_i}{\partial p_i^l}|_{m=\bar{m}} = -\sum_{l \in \mathcal{R}_i} \frac{\partial t_i^l}{\partial p_i^l}|_{m=\bar{m}} \tag{A.46}$$

and

$$\frac{\partial t_i^l}{\partial p_i^l}|_{m=\bar{m}} = \begin{cases} 0 & \text{if } |\mathcal{G}^l| = 2; \\ -2p^{*l}\left[\frac{x_i^*+x_j^*+x_k^*-c_l}{\gamma}\right] = 0 & \text{if } |\mathcal{G}^l| = 3; \\ -2P_{-i}^{*l}\left[\frac{\mathcal{E}_{-i}^{*l}+x_i^*}{\gamma}\right] = 0 & \text{if } |\mathcal{G}^l| > 3, \end{cases} \tag{A.47}$$

for any $l \in \mathbf{L}$ because of (A.41). Hence, $(x_1^*, x_2^*, \ldots, x_N^*, \lambda^{*l_1}, \lambda^{*l_2}, \ldots, \lambda^{*l_L})$ is a NE point of the game induced by the game form proposed in Sect. 3.3. □

Proof of Theorem 3.7. Since any NE of the game induced by the mechanism proposed in Sect. 3.3, (if such an equilibrium exists), results in a feasible allocation of Problem **Max**, (see Theorem 3.1), we restrict attention to the space $\mathcal{M}' = \mathcal{M}'_1 \times \mathcal{M}'_2 \cdots \mathcal{M}'_N$ of strategies that result in feasible allocations of Problem **Max**. Then, the users' utilities $\mathcal{V}_i(x_i, \mathbf{t}_i) = u_i(x_i) - \mathbf{t}_i$, $i = 1, 2, \ldots, N$, (where $\mathbf{t}_i$ is specified by the game form of Sect. 3.3) are quasi-concave in $m_i = (x_i, \mathrm{p}_i)$ and continuous in $m = (m_1, \ldots, m_N) = ((x_1, \mathrm{p}_1), (x_2, \mathrm{p}_2), \ldots, (x_N, \mathrm{p}_N))$. Furthermore, the message/strategy spaces $\mathcal{M}'_i$ are compact, convex and non-empty. Therefore, by Glicksberg's theorem [3], there exists a pure NE of the game $(\mathcal{M}, f, \mathcal{V}_i, i = 1, 2, \ldots, N)$ induced by the game form of Sect. 3.3.

Let m^* be a NE of this game. Then, for every user $i \in \mathcal{N}$,

$$\mathcal{V}_i(m^*) \geq \mathcal{V}_i(m_{-i}^*, m_i) \quad \text{for every } m_i \in \mathcal{M}_i. \tag{A.48}$$

That is,

$$u_i(x_i^*) - \sum_{l\in\mathcal{R}_i} t_i^{*l}(m^*) \geq u_i(x_i) - \sum_{l\in\mathcal{R}_i} t_i^l(m_{-i}^*, m_i) \;\; \forall\, m_i \in \mathcal{M}_i, \tag{A.49}$$

where

$$\sum_{l\in\mathcal{R}_i} t_i^{*l}(m^*) = \sum_{l\in\mathcal{R}_i} p^{*l}x_i^* + \sum_{\substack{l\in\mathcal{R}_i\\|\mathcal{G}^l|=3}} \Omega_i^{*l} + \sum_{\substack{l\in\mathcal{R}_i\\|\mathcal{G}^l|>3}} \Phi_i^{*l} + \sum_{j=1}^{r} \mathcal{Q}^{*l_j} 1\{i = k_{l_j}\}, \tag{A.50}$$

$\mathcal{Q}^{*l_j}$ is given by $\mathcal{Q}^{*\{l:|\mathcal{G}^l|=2\}}$ or $\mathcal{Q}^{*\{l:|\mathcal{G}^l|=3\}}$ and

$$\sum_{l\in\mathcal{R}_i} t_i^l(m_{-i}^*, m_i) = \sum_{\substack{l\in\mathcal{R}_i\\|\mathcal{G}^l|=2}} \Pi_2(m_{-i}^*, m_i) + \sum_{\substack{l\in\mathcal{R}_i\\|\mathcal{G}^l|=3}} \Pi_3(m_{-i}^*, m_i) + \sum_{\substack{l\in\mathcal{R}_i\\|\mathcal{G}^l|>3}} \Pi_{>3}(m_{-i}^*, m_i)$$
$$+ \sum_{j=1}^{r} \mathcal{Q}^{*l_j} 1\{i = k_{l_j}\}, \tag{A.51}$$

where

$$\Pi_2(m^*_{-i}, m_i) := p^{*l} x_i + \frac{(p_i^l - p^{*l})^2}{\alpha} - 2p^{*l}(p_i^l - p^{*l})\left(\frac{x_i + x_j^* - c_l}{\gamma}\right),$$

$$\Pi_3(m^*_{-i}, m_i) := p^{*l} x_i + (p_i^l - p^{*l})^2 + \Omega_i^{*l} - 2p^{*l}(p_i^l - p^{*l})\left(\frac{x_i + x_j^* + x_k^* - c_l}{\gamma}\right),$$

$$\Pi_{>3}(m^*_{-i}, m_i) := p^{*l} x_i + (p_i^l - p^{*l})^2 + \Phi_i^{*l} - 2p^{*l}(p_i^l - p^{*l})\left(\frac{x_i + \mathcal{E}_{-i}^{*l}}{\gamma}\right).$$

Since (A.49) holds for every feasible (x_i, P_i), setting $p_i^l = p^{*l}$ for every $l \in \mathcal{R}_i$ we obtain,

$$\mathcal{V}_i(x_i^*, \mathrm{P}^*) = u_i(x_i^*) - \sum_{l \in \mathcal{R}_i} p^{*l} x_i^* \geq \mathcal{V}_i(x_i, \mathrm{P}^*) = u_i(x_i) - \sum_{l \in \mathcal{R}_i} p^{*l} x_i \quad \text{(A.52)}$$

for every feasible x_i. Therefore, for every $i = 1, 2, \ldots, N$,

$$x_i^* = \arg\max_{x_i \in \mathbf{D}^*_{-i}} \left\{ u_i(x_i) - \sum_{l \in \mathcal{R}_i} p^{*l} x_i \right\} \quad \text{(A.53)}$$

where $\mathbf{D}^*_{-i} := \left\{ x_i : 0 \leq x_i \leq \min_{l \in \mathcal{R}_i} \{c_l - \sum_{\substack{j \in \mathcal{G}^l \\ j \neq i}} x_j^*\} \right\}$. Consequently, $(\mathrm{x}^*, \mathrm{p}^*)$ is a Walrasian equilibrium, therefore $(\mathrm{x}^*, \mathbf{t}^*)$ is Pareto optimal ([4] Chap. 15). □

Appendix B
Appendix for Multi-rate Multicast Service Provisioning

Proof of Theorem 5.2. We prove in Theorem 5.3 that any NE of the game induced by the mechanism of Sect. 5.3 (if such an equilibrium exists) results in a feasible allocation of Problem **Max.0**. Therefore, we restrict to the space

$$\mathcal{M} = \times_{G_i \in \mathcal{N}} \times_{j \in G_i} \mathcal{M}_{(j,G_i)} \tag{B.1}$$

of strategies that result in feasible allocations of problem **Max.0**. Then, the users' utilities

$$V_{(j,G_i)}(x_{(j,G_i)}, t_{(j,G_i)}) = U_{(j,G_i)}(x_{(j,G_i)}) - t_{(j,G_i)} \tag{B.2}$$

$(j, G_i) \in G_i$, $G_i \in \mathcal{N}$ (where $t_{(j,G_i)}$ is specified by the game form of Sect. 5.3) are concave in $\boldsymbol{m}_{(j,G_i)} = (x_{(j,G_i)}, \pi_{(j,G_i)})$ and continuous in $\boldsymbol{m} = (\boldsymbol{m}_{(j,G_i)}, (j, G_i) \in G_i, G_i \in \mathcal{N})$. Furthermore, the message spaces $\mathcal{M}_{(j,G_i)}$ are compact, convex and nonempty. Therefore, by Gliksberg's theorem, [5], there exists a pure NE of the game $(\mathcal{M}, f, V_{(j,G_i)}, (j, G_i) \in G_i, G_i \in \mathcal{N})$ induced by the game form of Sect. 5.3. □

Proof of Theorem 5.3. By the construction of the mechanism $x^*_{(j,G_i)} \geq 0$ for all (j, G_i), $G_i \in \mathcal{N}$. Suppose that $\boldsymbol{x}^*$ is such that the capacity constraint is violated at some link l and $x^*_{G_i}(l) > 0$. Consider an agent $(k, G_i) \in G_i^{\max}(l)$ whose index in $G_i^{\max}(l)$ is $(j, G_i^{\max}(l))$ and change his strategy to $x_{(k,G_i)} = 0$. Then

$$V_{(k,G_i)}(\boldsymbol{m}_{(k,G_i)}, \boldsymbol{m}^*_{-(k,G_i)}) > V_{(j,G_i)}(\boldsymbol{m}^*_{(k,G_i)}, \boldsymbol{m}^*_{-(k,G_i)}),$$

and this is in contradiction with the fact that $\boldsymbol{m}^*$ is a NE. Consequently, every NE results in a feasible allocation of problem **Max.0**. □

Proof of Lemma 5.4. We prove this lemma for **Case D, Part DI**. In a way similar to the following we can prove the assertion of the lemma for all other cases.

Case D (Part DI): Consider $G_i \in Q_l$, and $(j, G_i^{\max}(l)) \in G_i^{\max}(l)$.

A. Kakhbod, *Resource Allocation in Decentralized Systems with Strategic Agents*, Springer Theses, DOI: 10.1007/978-1-4614-6319-1,
© Springer Science+Business Media New York 2013

Since user $(j, G_i^{\max}(l))$ does not control $\Gamma_{(j,G_i)}$, then

$$\frac{\partial \Gamma_{(j,G_i^{\max}(l))}}{\partial \pi_{(j,G_i^{\max}(l))}} = \frac{\partial \Gamma_{(j,G_i^{\max}(l))}}{\partial x_{G_i}(l)} = 0.$$

Therefore, we must have

$$\left.\frac{\partial t^l_{(j,G_i^{\max}(l))}}{\partial \pi_{(j,G_i{}^{\max}(l))}}\right|_{m=m^*} = -2\frac{P^*_{-G_i{}^{\max}(l)}}{|G_i{}^{\max}(l)|}\left(\frac{\mathcal{E}^*_{-G_i{}^{\max}(l)} + x^*_{G_i}(l)}{\gamma}\right) + \frac{2}{|G_i{}^{\max}(l)|}\left(P^*_{G_i{}^{\max}(l)} - P^*_{-G_i{}^{\max}(l)}\right) = 0. \quad \text{(B.3)}$$

Define $\Delta_{(j,G_i{}^{\max}(l))}$ as follows,

$$\Delta_{(j,G_i{}^{\max}(l))} := -\frac{P^*_{-G_i{}^{\max}(l)}}{|G_i{}^{\max}(l)|}\left(\frac{\mathcal{E}^*_{-G_i{}^{\max}(l)} + x^*_{G_i}(l)}{\gamma}\right) + \frac{\left(P^*_{G_i{}^{\max}(l)} - P^*_{-G_i{}^{\max}(l)}\right)}{|G_i{}^{\max}(l)|}. \quad \text{(B.4)}$$

Summing over all the users in $G_i^{\max}(l)$ and using (B.3) we obtain

$$\sum_{(j,G_i^{\max})\in G_i{}^{\max}(l)} \Delta_{(j,G_i{}^{\max}(l))} = -P^*_{-G_i{}^{\max}(l)}\left(\frac{\mathcal{E}^*_{-G_i{}^{\max}(l)} + x^*_{G_i}(l)}{\gamma}\right) + \left(P^*_{G_i{}^{\max}(l)} - P^*_{-G_i{}^{\max}(l)}\right) = 0. \quad \text{(B.5)}$$

Moreover, summing over all $|Q_l|$ multicast groups and using (B.3)–(B.5) we get

$$\sum_{G_i\in Q_l}\sum_{(j,G_i^{\max}(l))\in G_i{}^{\max}(l)} \frac{\partial t^l_{(j,G_i^{\max}(l))}}{\partial \pi_{(j,G_i{}^{\max}(l))}} = \sum_{G_i\in Q_l}\sum_{(j,G_i^{\max}(l))\in G_i{}^{\max}(l)} \Delta_{(j,G_i{}^{\max}(l))} = 0. \quad \text{(B.6)}$$

Furthermore we note that

$$\sum_{G_i\in Q_l} P_{G_i{}^{\max}(l)} = \sum_{G_i\in Q_l} P_{-G_i{}^{\max}(l)}. \quad \text{(B.7)}$$

Equations (B.5)–(B.7) along with Theorem 5.3 and the fact that $P^*_{-G_i{}^{\max}(l)} \geq 0$ for every G_i, $G_i \in Q_l$, imply that

$$P^*_{-G_i{}^{\max}(l)}\left(\frac{\mathcal{E}^*_{-G_i{}^{\max}(l)} + x^*_{G_i}(l)}{\gamma}\right) = 0, \qquad \forall\, G_i \in Q_l. \quad \text{(B.8)}$$

From Eqs. (B.5) and (B.8) it follows that

$$P^*_{-G_i{}^{\max}(l)} = P^*_{G_i{}^{\max}(l)} =: P^*_{G^{\max}(l)}, \qquad \forall\, G_i \in Q_l. \tag{B.9}$$

Consequently,

$$P^*_{G^{\max}(l)} \left(\frac{\mathcal{E}^*_{-G_i{}^{\max}(l)} + x^*_{G_i}(l)}{\gamma} \right) = 0. \tag{B.10}$$

Equations (B.9) and (B.10) along with (5.24) give

$$\begin{aligned} \left. \frac{\partial t^l_{(j,G_i^{\max}(l))}}{\partial x_{G_i}(l)} \right|_{m=m^*} &= \pi^*_{(j+1,G_i{}^{\max}(l))} - 2\frac{P^*_{-G_i{}^{\max}(l)}}{\gamma |G_i{}^{\max}(l)|} \left(P^*_{G_i{}^{\max}(l)} - P^*_{-G_i{}^{\max}(l)} \right) \\ &= \pi^*_{(j+1,G_i{}^{\max}(l))}. \end{aligned} \tag{B.11}$$

□

Proof of Lemma 5.6. Equation (5.41) together with (5.42) and (5.43) imply that $\sum_{(j,G_i) \bigcup_{G_i \in \mathcal{N}} G_i} t^*_{(j,G_i)} = \sum_{l \in \mathbf{L}} \sum_{G_i \in Q_l} \sum_{j \in G_i} t^{*l}_{(j,G_i)} = 0$. Thus, the mechanism is budget balanced at allocations corresponding to NE. Now, we prove that the proposed mechanism is also budget balanced off equilibrium.

For that matter, we first consider links $l \in \mathbf{L}$ where $|Q_l| > 3$. Thus we have,

$$\begin{aligned} \sum_{(j,G_i) \in G_i(l)} t^l_{(j,G_i)} &= P_{G_i{}^{\max}(l)} x_{G_i}(l) + \left(P_{G_i{}^{\max}(l)} - P_{-G_i{}^{\max}(l)} \right)^2 + \Gamma^l_{G_i} \\ &\quad - 2P_{-G_i{}^{\max}(l)} \left(P_{G_i{}^{\max}(l)} - P_{-G_i{}^{\max}(l)} \right) \left(\frac{\mathcal{E}_{-G_i{}^{\max}(l)} + x_{G_i}(l)}{\gamma} \right). \end{aligned} \tag{B.12}$$

Furthermore, by little algebra, we can show that for every $l \in \mathbf{L}$ where $|Q^l| > 3$ the following equalities hold,

$$\sum_{G_i \in Q_l} P^2_{G_i^{\max}(l)} = \sum_{G_i \in Q_l} \left[\frac{\sum_{\substack{G_j \in Q_l \\ G_j \neq G_i}} P^2_{G_i^{\max}(l)}}{|Q_l| - 1} \right],$$

$$\begin{aligned} &\sum_{G_i \in Q_l} \left[2P_{G_i^{\max}(l)} P_{-G_i^{\max}(l)} + 2P_{-G_i^{\max}(l)} P_{G_i^{\max}(l)} \frac{x_{G_i}(l)}{\gamma} \right] \\ &= \sum_{G_i \in Q_l} \left[\frac{\sum_{\substack{G_j \in Q_l \\ G_j \neq G_i}} \sum_{\substack{G_k \in Q_l \\ G_k \neq G_i, G_j}} \left(2P_{G_j^{\max}(l)} P_{G_k^{\max}(l)} (1 + \frac{x_{G_j}(l)}{\gamma}) \right)}{(|Q_l| - 1)(|Q_l| - 2)} \right], \end{aligned}$$

$$\sum_{G_i \in Q_l} P_{-G_i^{\max}(l)} P_{G_i^{\max}(l)} \frac{\mathcal{E}_{-G_i^{\max}(l)}}{\gamma}$$
$$= \frac{\sum_{G_i \in Q_l} \sum_{\substack{G_j \in Q_l \\ G_j \neq G_i}} \sum_{\substack{G_k \in Q_l \\ G_k \neq G_i, G_j}} 2 P_{G_k^{\max}(l)} P_{G_j^{\max}(l)} \mathcal{E}_{G_k^{\max}(l)}}{\gamma(|Q_l|-1)^2(|Q_l|-2)}$$
$$+ \frac{\sum_{G_i \in Q_l} \sum_{\substack{G_j \in Q_l \\ G_j \neq G_i}} \sum_{\substack{G_k \in Q_l \\ G_k \neq G_i, G_j}} \sum_{\substack{G_r \in Q_l \\ G_r \neq G_i, G_j, G_k}} 2 P_{G_k^{\max}(l)} P_{G_j^{\max}(l)} \mathcal{E}_{G_r^{\max}(l)}}{\gamma(|Q_l|-1)^2(|Q_l|-3)}$$

$$\sum_{G_i \in Q_l} P^2_{-G_i^{\max}(l)} \frac{x_{G_i}(l)}{\gamma} = \sum_{G_i \in Q_l} \left[\frac{\sum_{\substack{G_j \in Q_l \\ G_j \neq G_i}} \sum_{\substack{G_k \in Q_l \\ G_k \neq G_i, G_j}} \sum_{\substack{G_r \in Q_l \\ G_r \neq G_i, G_j, G_k}} x_{G_j}(l) P_{G_r^{\max}(l)}}{\gamma(|Q_l|-1)^2(|Q_l|-3)} + \frac{\sum_{\substack{G_j \in Q_l \\ G_j \neq G_i}} \sum_{\substack{G_k \in Q_l \\ G_k \neq G_i, G_j}} x_{G_j}(l) P_{G_k^{\max}(l)}}{\gamma(|Q_l|-1)^2(|Q_l|-2)} \right] \tag{B.13}$$

Using Eqs. (5.31) and (B.12) –(B.13) we obtain

$$\sum_{G_i \in Q_l} \left[\left(P_{G_i{}^{\max}(l)} - P_{-G_i{}^{\max}(l)} \right)^2 \right]$$
$$- \sum_{G \in Q_l} \left[2 P_{-G_i{}^{\max}(l)} \left(P_{G_i{}^{\max}(l)} - P_{-G_i{}^{\max}(l)} \right) \left(\frac{\mathcal{E}_{-G_i{}^{\max}(l)} + x_{G_i}(l)}{\gamma} \right) \right]$$
$$+ \sum_{G_i \in Q_l} \Gamma^l_{G_i} = 0. \tag{B.14}$$

Next we consider all links $l \in \mathbf{L}$ where $|Q_l| \leq 3$; let these link be $l_1, l_2, \ldots, l_r$. Then, by using (B.14) and the specification of the tax function for the links $l_1, l_2, \ldots, l_r$ (cf. Sect. 5.3, cases B and C) we obtain

$$\sum_{(j,G_i) \in \bigcup_{G_i \in \mathcal{N}} G_i} t_{(j,G_i)} = \sum_{l \in \mathbf{L}} \sum_{G_i \in Q_l} \sum_{(j,G_i) \in G_i(l)} t^l_{(j,G_i)}$$
$$= \sum_{l \in \mathbf{L}: |Q_l|=2} \sum_{G_i \in Q_l} \sum_{(j,G_i) \in G_i{}^{\max}(l)} t^l_{(j,G_i)}$$
$$+ \sum_{l \in \mathbf{L}: |Q_l|=3} \sum_{G_i \in Q_l} \sum_{(j,G_i) \in G_i{}^{\max}(l)} t^l_{(j,G_i)}$$

$$+ \sum_{l \in \mathbf{L}:|Q_l|>3} \sum_{G_i \in Q_l} \sum_{(j,G_i) \in G_i{}^{\max}(l)} t^l_{(j,G_i)} + \sum_{j=1}^{r} \mathcal{S}^{l_j}$$
$$= 0. \tag{B.15}$$

The last equality in (B.15) is true for the following reason. By Eq. (B.14) the third sum on the right hand side of the second equality in (B.15) is equal to zero. The sum of the three remaining terms is also equal to zero because of Eqs. (5.16)–(5.35). □

Proof of Theorem 5.7. We need to show that

$$V_{(j,G_i)}(\boldsymbol{m}^*) = \left[U_{(j,G_i)}(x_{(j,G_i)}) - \sum_{l \in \mathcal{R}_{(j,G_i)}} t^l_{(j,G_i)} \right]_{\boldsymbol{m}=\boldsymbol{m}^*} \geq 0,$$

for every (j, G_i), $G_i \in \mathcal{N}$. By the property of NE, it follows that

$$V_{(j,G_i)}\left(\boldsymbol{m}^*\right) \geq V_{(j,G_i)}\left(\boldsymbol{m}^*_{-(j,G_i)}, \boldsymbol{m}_{(j,G_i)}\right). \tag{B.16}$$

Consequently, it is sufficient to find a $\boldsymbol{m}_{(j,G_i)} \in \mathcal{M}_i$ so that $V_{(j,G_i)}\big(\boldsymbol{m}^*_{-(j,G_i)}, \boldsymbol{m}_{(j,G_i)}\big) \geq 0$. Set $x_{(j,G_i)}$ equal to 0. We separately examine different cases, as follows.

- If $x^*_{G_i}(l) > 0$ then, $t^l_{(j,G_i)}|_{(\boldsymbol{m}^*_{-(j,G_i)}, \boldsymbol{m}_{(j,G_i)})} = 0$ because $j \notin G_i{}^{\max}(l)$.
- If $x^*_{G_i}(l) = 0$, then in accordance to the possible cases we define,

$$\pi^l_{(j,G_i)} := \begin{cases} \pi^{*l}_{(j,G_i)}, & \text{for Case B, Part BI;} \\ P^*_{G^{\max}(l)}, & \text{for Case B, Part BII;} \\ \pi^{*l}_{(j,G_i)}, & \text{for Case C, Part CI;} \\ P^*_{G^{\max}(l)}, & \text{for Case C, Part CII;} \\ \varpi^*_{DI}(l), & \text{for Case D, Part DI;} \\ \varpi^*_{DII}(l), & \text{for Case D, Part DII.} \end{cases} \tag{B.17}$$

where

$$\varpi^*_{DI}(l) := \frac{1}{|G_i^{\max}(l)|} \Bigg[P^*_{G^{\max}(l)} - \sum_{\substack{j \in G_i^{\max} \\ j \neq i}} \pi^*_{j,G_i^{\max}} + \frac{\mathcal{E}^*_{-G_i{}^{\max}(l)}}{\gamma}$$
$$+ \sqrt{\left[P^*_{G^{\max}(l)} \frac{\mathcal{E}^*_{-G_i{}^{\max}(l)}}{\gamma} \right]^2 + \frac{P^*_{G^{\max}(l)} \sum_{G_j, G_j \in Q_l, G_j \neq G_i} x^*_{G_j}(l)}{|Q_l| - 1}} \Bigg],$$

$$\varpi^*_{DII}(l) := P^*_{G^{\max}(l)}\left[1 + \frac{\mathcal{E}^*_{-G_i{}^{\max}(l)}}{\gamma}\right] + \sqrt{\left[P^*_{G^{\max}(l)}\frac{\mathcal{E}^*_{-G_i{}^{\max}(l)}}{\gamma}\right]^2 + \frac{P^*_{G^{\max}(l)}\sum_{\substack{G_j, G_j \in Q_l \\ G_j \neq G_i}} x^*_{G_j}(l)}{|Q_l| - 1}}.$$

We can[1] show that $t^l_{(j,G_i)}$ for every $l \in \mathcal{R}_{(j,G_i)}$ is equal to zero at $\boldsymbol{m}_{(j,G_i)} = (0, \pi^{l_1}_{(j,G_i)}, \ldots, \pi^{l_{|\mathcal{R}_{(j,G_i)}|}}_{(j,G_i)})$ when $\pi^{l_k}_{(j,G_i)}$, $1 \leq k \leq |\mathcal{R}_{(j,G_i)}|$ is defined by (B.17).

In the other hand, by $\boldsymbol{m}_{(j,G_i)}$ where its arguments are defined in the above, we obtain

$$\begin{aligned} V_{(j,G_i)}\left(\boldsymbol{m}^*_{-(j,G_i)}, \boldsymbol{m}_{(j,G_i)}\right) &= U_{(j,G_i)}(0) - \sum_{l \in \mathcal{R}_{(j,G_i)}} t^l_{(j,G_i)}\left(\boldsymbol{m}^*_{-(j,G_i)}, \boldsymbol{m}_{(j,G_i)}\right) \\ &= U_{(j,G_i)}(0) \\ &= 0, \end{aligned} \tag{B.18}$$

when $(j, G_i) \neq k_{l_1}, k_{l_2}, \ldots, k_{l_r}$.
When $(j, G_i) = k_{l_q}, q = 1, 2, \ldots, r$,

$$\begin{aligned} V_{(j,G_i)}\left(\boldsymbol{m}^*_{-(j,G_i)}, \boldsymbol{m}_{(j,G_i)}\right) &= U_{(j,G_i)}(0) - \sum_{l \in \mathcal{R}_{(j,G_i)}} t^l_{(j,G_i)}\left(\boldsymbol{m}^*_{-(j,G_i)}, \boldsymbol{m}_{(j,G_i)}\right) - \mathcal{S}^{*l_q} \\ &= -\mathcal{S}^{*l_q} \\ &\geq 0, \end{aligned} \tag{B.19}$$

Combining (B.16), (B.18) and (B.19) we obtain

$$V_{(j,G_i)}(x^*_{(j,G_i)}, \boldsymbol{t}^*) \geq V_{(j,G_i)}(x, \boldsymbol{t})\Big|_{\boldsymbol{m}=(\boldsymbol{m}_{(j,G_i)}, \boldsymbol{m}^*_{-(j,G_i)})} \geq 0 \tag{B.20}$$

□

Proof of Theorem 5.8. Let $\boldsymbol{m}^*$ be an arbitrary *NE* of the game $(\mathcal{M}, f, V)$ induced by the proposed game form. Consider problem **Max.1**, since the functions $U_{(j,G_i)}$, $j \in G_i$, $G_i \in \mathcal{N}$, are concave and differentiable and the constraints are linear, Slater's condition [2] is satisfied, the duality gap is equal to zero, and Karush Kuhn Tucker (KKT) conditions are necessary and sufficient to guarantee the optimality of any allocation $\boldsymbol{x}$ that satisfies them. Let λ^l be the Lagrange multiplier corresponding to the capacity constraint for link l and ν_i be the Lagrange multiplier corresponding to

[1] Since γ is sufficiently large then it is guaranteed that ϖ_{DI} and ϖ_{DII} are positive.

the demand constraint. The Lagrangian for problem **Max.1** is

$$
\begin{aligned}
\mathcal{L}x(x,\lambda,\nu) = & \sum_{G_i \in \mathcal{N}} \sum_{(j,G_i)\in G_i} U_{(j,G_i)}(x_{(j,G_i)}) \\
& - \sum_{l\in \mathbf{L}} \sum_{e(l)\in E(l)} \lambda_{e(l)} \left[\sum_{G_i \in Q_l} x_{(j,G_i)} \mathbb{I}\{(j,G_i)\in G_i(l)\} - c_l \right] \\
& + \sum_{G_i \in \mathcal{N}} \sum_{(j,G_i)\in G_i} \nu_{(j,G_i)} x_{(j,G_i)}
\end{aligned} \tag{B.21}
$$

and the KKT conditions are:

$$
\begin{aligned}
\frac{\partial \mathcal{L}(\mathbf{x}^*,\lambda^*,\nu^*)}{\partial x_{(j,G_i)}} = & \frac{\partial U_{(j,G_i)}(x^*_{(j,G_i)})}{\partial x_{(j,G_i)}} \\
& - \sum_{l\in \mathcal{R}_{(j,G_i)}} \sum_{e(l,(j,G_i))\in E(l,(j,G_i))} \lambda^*_{e(l,(j,G_i))} + \nu^*_{(j,G_i)} = 0
\end{aligned}
$$

$$
\lambda^*_{e(l)} \left[\sum_{G_i \in Q_l} x^*_{(j,G_i)} \mathbb{I}\{(j,G_i)\in G_i(l)\} - c_l \right] = 0, \quad \forall\, l \in \mathbf{L} \tag{B.22}
$$

$$
\nu^*_{(j,G_i)} x^*_{(j,G_i)} = 0 \quad \forall G_i \in \mathcal{N} \text{ and } j \in G_i. \tag{B.23}
$$

Now, define

$$
\lambda^{*l}_{(j,G_i)} := \sum_{e^{\max}(l,(j,G_i))\in E^{\max}(l,(j,G_i))} \lambda^*_{e(l,(j,G_i))} \quad \forall\, l \in \mathbf{L},\ G_i \in Q_l,\ j \in {G_i}^{\max}(l), \tag{B.24}
$$

where, $E^{\max}(l)$ is a subset of equations, $e^{\max}(l)$ of (5.7), such that every element $x_{(k,G_s)} \in e^{\max}(l)$ is equal to $x_{G_s}(l)$, and accordingly, we can define $E^{\max}(l,(j,G_i))$ and $e^{\max}(l,(j,G_i))$.

Furthermore, (B.22) implies the following

$$
\forall\, l \in \mathbf{L} \text{ and } G_i \in Q_l,\ j \in G_i(l),\ \text{if } x_{(j,G_i)} < x_{G_i}(l) \text{ then } \lambda^*_{e(l,(j,G_j))} = 0. \tag{B.25}
$$

Since $\boldsymbol{m}^*$ is a NE then for every user $(j,G_i), G_i \in \mathcal{N}, j \in G_i$, there exists at least a link in $\mathcal{R}_{(j,G_i)}$ such that $x_{(j,G_i)} = x_{G_i}(l)$. Now, by using (B.24) and (B.25) we can reformulate the KKT constraints as follows, suppose that at link $l \in \mathcal{R}_{(j,G_i)}, x_{(j,G_i)} = x_{G_i}(l)$, then

$$
\frac{\partial \mathcal{L}(\mathbf{x}^*,\lambda^*,\nu^*)}{\partial x^*_{G_i}(l)} = \frac{\partial U_{(j,G_i)}(x^*_{G_i}(l))}{\partial x_{G_i}(l)} - \sum_{l\in \mathcal{R}^{\max}_{(j,G_i)}} \lambda^{*l}_{(j,G_i)} + \nu^*_{(j,G_i)} = 0 \tag{B.26}
$$

$$\lambda^{*l}\left[\sum_{G_i\in Q_l} x^*_{G_i}(l) - c_l\right] = 0, \quad \forall l \in \mathbf{L} \tag{B.27}$$

$$\nu^*_{(j,G_i)} x^*_{G_i}(l) = 0 \quad \forall G_i \in \mathcal{N} \text{ and } j \in G_i. \tag{B.28}$$

where $\lambda^{*l} := \sum_{j\in G_i{}^{\max}(l)} \lambda^{*l}_{(j,G_i)}$ for every $G_i \in Q_l$.

Because of the characteristics of problem **Max.1**, KKT conditions are necessary and sufficient for any optimal solution of **Max.1**. Therefore, to show that any arbitrary NE $\boldsymbol{m}^*$ of the specified game, induced from the game form presented in Sect. 5.3, is correspondent to an optimal solution, it is enough to find ν^*_i, λ^{l*}, and $\lambda^{*l}_{(j,G_i)}$, for every $G_i \in \mathcal{N}$, $j \in G_i$, $l \in \mathbf{L}$, appropriately, such that Eqs. (B.26), (B.27) and (B.28) are satisfied. If we set $\nu^*_{(j,G_i)}$, $G_i \in \mathcal{N}$, $j \in G_i$, equal to zero, then (B.28) is satisfied. In addition, if we set $\lambda^{l*} = P^*_{G^{\max}(l)}$, $l \in \mathbf{L}$ and $\lambda^{*l}_{(j,G_i)}$ equal to (5.39, then the correctness of (B.27) and (B.28) will be implied from (5.38) and (5.40), respectively. Furthermore, by the construction of the game form $\sum_{G_i\in\mathcal{N}}\sum_{j\in G_i}\sum_{l\in\mathcal{R}_{(j,G_i)}} t^{*l}_{(j,G_i)}$ is equal to zero. Consequently, the NE $\boldsymbol{m}^*$ results in an optimal solution of problem **Max.0**. Since the NE $\boldsymbol{m}^*$ was arbitrary chosen, every NE $\boldsymbol{m}^*$ of the game induced by the game form proposed in Sect. 5.3 results in an optimal solution of problem **Max.0**. □

References

1. Bracwell R (1999) The fourier transform and its applications. McGraw-Hill, New York
2. Boyd S, Vandenberghe L (2004) Convex optimization. Cambridge University Press, Cambridge
3. Myerson R (1981) Optimal auction design. Math Oper Res 6(1):58–73
4. Mas-Colell A, Whinston MD, Green JR (2005) Microeconomic theory. Oxford University Press, New York
5. Fudenberg D, Tirole J (1991) Game theory. MIT Press, Cambridge

Printed in the United States
By Bookmasters